THE FOURTH INDUSTRIAL REVOLUTION: HOPES AND FEARS

DR. NASSER AFIFY

2018

TABLE OF CONTENTS

INTRODUCTION...	1
1- DEFINTION OF THE INDUSTRIAL REVOLUTION...	10
2- THE EMERGENCE OF THE FOURTH INDUSTRIAL REVOLUTION	37
3- BENEFITS AND RISKS OF THE FOURTH INDUSTRIAL REVOLUTION	58
4- THE IMPACT OF FOURTH INDUSTRIAL ROVUTION ON LABOR	75
5- THE IMPACT OF FOURTH INDUSTRIAL REVOLUTION ON HIGHER EDUCATION	101
6- THE IMPACT OF FOURTH INDUSTRIAL REVOLUTION ON HEALTHCARE	121
7- THE IMPACT OF FOURTH INDUSTRIAL REVOLUTION ON AGRICULTURE	126
CONCLUSION..	137
REFERENCES..	140

INTRODUCTION

We stand on the brink of a technological revolution that will fundamentally alter the way we live, work, and relate to one another. In its scale, scope, and complexity, the transformation will be unlike anything humankind has experienced before. We do not yet know just how it will unfold, but one thing is clear: the response to it must be integrated and comprehensive, involving all stakeholders of the global polity, from the public and private sectors to academia and civil society.

The First Industrial Revolution used water and steam power to mechanize production. The Second used electric power to create mass production. The Third used electronics and information technology to automate production. Now a Fourth Industrial Revolution is building on the Third, the digital revolution that has been occurring since the middle of the last century. It is characterized by a fusion of technologies that is blurring the lines between the physical, digital, and biological spheres.

There are three reasons why today's transformations represent not merely a prolongation of the Third Industrial Revolution but rather the arrival of a Fourth and distinct one: velocity, scope, and systems impact. The speed of current breakthroughs has no historical precedent. When compared with previous industrial revolutions, the Fourth is evolving at an exponential rather than a linear pace. Moreover, it is disrupting almost every industry in every country. And the breadth and depth of these changes herald the transformation of entire systems of production, management, and governance.

The possibilities of billions of people connected by mobile devices, with unprecedented processing power, storage capacity, and access to knowledge, are unlimited. And these possibilities will be multiplied by emerging technology breakthroughs in fields such as artificial intelligence, robotics, the Internet of Things, autonomous vehicles, 3-D printing, nanotechnology, biotechnology, materials science, energy storage, and quantum computing.

Like the revolutions that preceded it, the Fourth Industrial Revolution has the potential to raise global income levels and improve the quality of life for populations around the world. To date, those who have gained the most from it have been consumers able to afford and access the digital world; technology has made possible new products and services that increase the efficiency and pleasure of our personal lives. Ordering a cab, booking a flight, buying a product, making a payment, listening to music, watching a film, or playing a game—any of these can now be done remotely.

In the future, technological innovation will also lead to a supply-side miracle, with long-term gains in efficiency and productivity. Transportation and communication costs will drop, logistics and global supply chains will become more effective, and the cost of trade will diminish, all of which will open new markets and drive economic growth.

In addition to being a key economic concern, inequality represents the greatest societal concern associated with the Fourth Industrial Revolution. The largest beneficiaries of innovation tend to be the providers of intellectual and physical capital—the innovators, shareholders, and investors—which explain the rising gap in wealth between those dependent on capital versus labor. Technology is therefore one of the main reasons why incomes have stagnated, or even decreased, for a majority of the population in high-income countries: the demand for highly skilled workers has increased while the demand for workers with less education and lower skills has decreased. The result is a job market with a strong demand at the high and low ends, but a hollowing out of the middle.

This helps explain why so many workers are disillusioned and fearful that their own real incomes and those of their children will continue to stagnate. It also helps explain why middle classes around the world are increasingly experiencing a pervasive sense of dissatisfaction and unfairness. A winner-takes-all economy that offers only limited access to the middle class is a recipe for democratic malaise and dereliction.

Discontent can also be fueled by the pervasiveness of digital technologies and the dynamics of information sharing typified by social media. More than 30 percent of the global population now uses social media platforms to connect, learn, and share information. In an ideal world, these interactions would provide an opportunity for cross-cultural understanding and cohesion. However, they can also create and propagate unrealistic expectations as to what constitutes success for an individual or a group, as well as offer opportunities for extreme ideas and ideologies to spread.

On the supply side, many industries are seeing the introduction of new technologies that create entirely new ways of serving existing needs and significantly disrupt existing industry value chains. Disruption is also flowing from agile, innovative competitors who, thanks to access to global digital platforms for research, development, marketing, sales, and distribution, can oust well-established incumbents faster than ever by improving the quality, speed, or price at which value is delivered.

Major shifts on the demand side are also occurring, as growing transparency, consumer engagement, and new patterns of consumer behavior (increasingly built upon access to mobile networks and data) force companies to adapt the way they design, market, and deliver products and services.

A key trend is the development of technology-enabled platforms that combine both demand and supply to disrupt existing industry structures, such as those we see within the "sharing" or "on demand" economy. These technology platforms, rendered easy to use by the smartphone, convene people, assets, and data—thus creating entirely new ways of consuming goods and services in the process. In addition, they lower the barriers for businesses and individuals to create wealth, altering the personal and professional environments of workers. These new platform businesses are rapidly multiplying into many new services, ranging from laundry to shopping, from chores to parking, from massages to travel.

On the whole, there are four main effects that the Fourth Industrial Revolution has on business—on customer expectations, on product enhancement, on collaborative innovation, and on organizational forms. Whether consumers or businesses, customers are increasingly at the epicenter of the economy, which is all about improving how customers are served. Physical products and services, moreover, can now be enhanced with digital capabilities that increase their value. New technologies make assets more durable and resilient, while data and analytics are transforming how they are maintained. A world of customer experiences, data-based services, and asset performance through analytics, meanwhile, requires new forms of collaboration, particularly given the speed at which innovation and disruption are taking place. And the emergence of global platforms and other new business models, finally, means that talent, culture, and organizational forms will have to be rethought.

Overall, the inexorable shift from simple digitization (the Third Industrial Revolution) to innovation based on combinations of technologies (the Fourth Industrial Revolution) is forcing companies to reexamine the way they do business. The bottom line, however, is the same: business leaders and senior executives need to understand their changing environment, challenge the assumptions of their operating teams, and relentlessly and continuously innovate.

As the physical, digital, and biological worlds continue to converge, new technologies and platforms will increasingly enable citizens to engage with governments, voice their opinions, coordinate their efforts, and even circumvent the supervision of public authorities. Simultaneously, governments will gain new technological powers to increase their control over populations, based on pervasive surveillance systems and the ability to control digital infrastructure. On the whole, however, governments will increasingly face pressure to change their current approach to public engagement and policymaking, as their central role of conducting policy diminishes owing to new sources of competition and the redistribution and decentralization of power that new technologies make possible.

Ultimately, the ability of government systems and public authorities to adapt will determine their survival. If they prove capable of embracing a world of disruptive change, subjecting their structures to the levels of transparency and efficiency that will enable them to maintain their competitive edge, they will endure. If they cannot evolve, they will face increasing trouble.

This will be particularly true in the realm of regulation. Current systems of public policy and decision-making evolved alongside the Second Industrial Revolution, when decision-makers had time to study a specific issue and develop the necessary response or appropriate regulatory framework. The whole process was designed to be linear and mechanistic, following a strict "top down" approach.

But such an approach is no longer feasible. Given the Fourth Industrial Revolution's rapid pace of change and broad impacts, legislators and regulators are being challenged to an unprecedented degree and for the most part are proving unable to cope.

The Fourth Industrial Revolution will also profoundly impact the nature of national and international security, affecting both the probability and the nature of conflict. The history of warfare and international security is the history of technological innovation, and today is no exception. Modern conflicts involving states are increasingly "hybrid" in nature, combining traditional battlefield techniques with elements previously associated with nonstate actors. The distinction between war and peace, combatant and noncombatant, and even violence and nonviolence is becoming uncomfortably blurry.

As this process takes place and new technologies such as autonomous or biological weapons become easier to use, individuals and small groups will increasingly join states in being capable of causing mass harm. This new vulnerability will lead to new fears. But at the same time, advances in technology will create the potential to reduce the scale or impact of violence, through the development of new modes of protection, for example, or greater precision in targeting.

The Fourth Industrial Revolution, finally, will change not only what we do but also who we are. It will affect our identity and all the issues associated with it: our sense of privacy, our notions of ownership, our consumption patterns, the time we devote to work and leisure, and how we develop our careers, cultivate our skills, meet people, and nurture relationships. It is already changing our health and leading to a "quantified" self, and sooner than we think it may lead to human augmentation. The list is endless because it is bound only by our imagination.

1- THE DEFINITION OF THE INDUSTRIAL REVOLUTION

The industrial revolution, in general, can be defined as the change in social and economic organization resulting from the replacement of hand tools by machine and power tools and the development of large-scale industrial production.

It is also a rapid major change in an economy (as in England in the late 18th century) marked by the general introduction of power-driven machinery or by an important change in the prevailing types and methods of use of such machines.

And it is the period of time during which work began to be done more by machines in factories than by hand at home. A period in which the development of machinery leads to major changes in agriculture, industry, transportation, and social conditions, esp. the Industrial Revolution in England in the 18th century.

The industrial revolution also is the rapid development of industry that occurred in Britain in the late 18th and 19th centuries, brought about by the introduction of machinery. It was characterized by the use of steam power, the growth of factories, and the mass production of manufactured goods.

Other definitions include the totality of the changes in economic and social organization that began about 1760 in England and later in other countries, characterized chiefly by the replacement of hand tools with power-driven machines, as the power loom and the steam engine, and by the concentration of industry in large establishments.

It is the rapid industrial growth that began in England during the middle of the eighteenth century and then spread over the next 50 years to many other countries, including the United States. The revolution depended on devices such as the steam engine, which were invented at a rapidly increasing rate during the period. The Industrial Revolution brought on a rapid concentration of people in cities and changed the nature of work for many people.

Industrial Revolution, in modern history, is the process of change from an agrarian and handicraft economy to one dominated by industry and machine manufacturing. This process began in Britain in the 18th century and from there spread to other parts of the world. Although used earlier by French writers, the term Industrial Revolution was first popularized by the English economic historian Arnold

Toynbee (1852–83) to describe Britain's economic development from 1760 to 1840.

The Industrial Revolution, the period in which agrarian and handicraft economies shifted rapidly to industrial and machine-manufacturing-dominated ones, began in the United Kingdom in the 18th century and later spread throughout many other parts of the world. This economic transformation changed not only how work was done and goods were produced, but it also altered how people related both to one another and to the planet at large.

This wholesale change in societal organization continues today, and it has produced several effects that have rippled throughout Earth's political, ecological, and cultural spheres. The following list describes some of the great benefits as well as some of the significant shortcomings associated with the Industrial Revolution.

Factories and the machines that they housed began to produce items faster and cheaper than could be made by hand. As the supply of various items rose, their cost to the consumer declined (see supply and demand). Shoes, clothing, household goods, tools, and other items that enhance people's quality of life became more common and less expensive.

Foreign markets also were created for these goods, and the balance of trade shifted in favor of the producer—which brought increased wealth to the companies that produced these goods and added tax revenue to government coffers. However, it also contributed to the wealth inequality between goods-producing and goods-consuming countries.

The rapid production of hand tools and other useful items led to the development of new types of tools and vehicles to carry goods and people from one place to another. The growth of road and rail transportation and the invention of the telegraph (and its associated infrastructure of telegraph—and later telephone and fiber optic—lines) meant that word of advances in manufacturing, agricultural harvesting, energy production, and medical techniques could be communicated between interested parties quickly.

Labor-saving machines such as the spinning jenny (a multiple-spindle machine for spinning wool or cotton) and other inventions, especially those driven by electricity (such as home appliances and refrigeration) and fossil fuels (such as automobiles and other fuel-powered vehicles), are also well-known products of the Industrial Revolution.

The Industrial Revolution was the engine behind various advances in medicine. Industrialization allowed medical instruments (such as scalpels, microscope lenses, test tubes, and other equipment) to be produced more quickly. Using machine manufacturing, refinements to these instruments could more efficiently roll out to the physicians that needed them. As communication between physicians in different areas improved, the details behind new cures and treatments for disease could be dispersed quickly, resulting in better care.

Mass production lowered the costs of much-needed tools, clothes, and other household items for the common people, which allowed them to save money for other things and build personal wealth. In addition, as new manufacturing machines were invented and new factories were built, new employment opportunities arose. No longer was the average person so closely tied to land-related concerns (such as being dependent upon the wages farm labor could provide or the plant and animal products farms could produce).

Industrialization reduced the emphasis on landownership as the chief source of personal wealth. The rising demand for manufactured goods meant that average people could make their fortunes in cities as factory employees and as employees of businesses that supported the factories, which paid better wages than farm-related positions. Generally speaking, people could save some portion of their wages, and many had the opportunity to invest in profitable businesses, thereby growing their family "nest eggs." The subsequent growth of the middle class in the United Kingdom and other industrializing societies meant that it was making inroads into the pool of economic power held by the aristocracy. Their greater buying power and importance in society led to changes in laws that were updated to better handle the demands of an industrialized society.

As industrialization progressed, more and more rural folk flocked to the cities in search of better pay in the factories. To increase the factories' overall efficiency and to take advantage of new opportunities in the market, factory workers were trained to perform specialized tasks. Factory owners divided their workers into different groups, each group focusing on a specific task. Some groups secured and transported to the factories raw materials (namely iron, coal, and steel) used in mass production of goods, while other groups operated different machines.

As the factories grew and workers became more specialized, additional teachers and trainers were needed to pass on specialized skills. In addition, the housing, transportation, and recreational needs of factory workers resulted in the rapid expansion of cities and towns. Governmental bureaucracies grew to support these, and new specialized departments were created to handle traffic, sanitation, taxation, and other services. Other businesses within the towns also became more specialized as more builders, physicians, lawyers, and other workers were added to handle the various needs of the new residents.

The promise of better wages attracted migrants to cities and industrial towns that were ill-prepared to handle them. Although initial housing shortages in many areas eventually gave way to construction booms and the development of modern buildings, cramped shantytowns made up of shacks and other forms of poor-quality housing appeared first.

Local sewerage and sanitation systems were overwhelmed by the sudden influx of people, and drinking water was often contaminated. People living in such close proximity, fatigued by poor working conditions, and drinking unsafe water presented ideal conditions for outbreaks of typhus, cholera, smallpox, tuberculosis, and other infectious diseases.

The need to treat these and other diseases in urban areas spurred medical advances and the development of modern building codes, health laws, and urban planning in many industrialized cities.

With relatively few exceptions, the world's modern environmental problems began or were greatly exacerbated by the Industrial Revolution. To fuel the factories and to sustain the output of each and every type of manufactured good, natural resources (water, trees, soil, rocks and minerals, wild and domesticated animals, etc.) were transformed, which reduced the planet's stock of valuable natural capital. The global challenges of widespread water and air pollution, reductions in biodiversity, destruction of wildlife habitat, and even global warming can be traced back to this moment in human history.

The more countries industrialize in pursuit of their own wealth, the greater this ecological transformation becomes. For example, atmospheric carbon dioxide, a primary driver of global warming, existed in concentrations of 275 to 290 parts per million by volume (ppmv) before 1750 and increased to more than 400 ppmv by 2017.

In addition, human beings use more than 40% of Earth's land-based net primary production, a measure of the rate at which plants convert solar energy into food and growth. As the world's human population continues to grow and more and more people strive for the material benefits promised by the Industrial Revolution, more and more of Earth's resources are appropriated for human use, leaving a dwindling stock for the plants and animals upon whose ecosystem services (clean air, clean water, etc.) the biosphere depends.

When factories sprung up in the cities and industrial towns, their owners prized production and profit over all else. Worker safety and wages were less important. Factory workers earned greater wages compared with agricultural workers, but this often came at the expense of time and less than ideal working conditions. Factory workers often labored 14–16 hours per day six days per week. Men's meager wages were often more than twice those of women.

The wages earned by children who worked to supplement family income were even lower. The various machines in the factory were often dirty, expelling smoke and soot, and unsafe, both of which contributed to accidents that resulted in worker injuries and deaths. The rise of labor unions, however, which began as a reaction to child labor, made factory work less grueling and less dangerous.

During the first half of the 20th century, child labor was sharply curtailed, the workday was reduced substantially, and government safety standards were rolled out to protect the workers' health and well-being.

As more cheap labor-saving devices become available, people performed less strenuous physical activity. While grueling farm-related labor was made far easier, and in many cases far safer, by replacing animal power and human power with tractors and other specialized vehicles to till the soil and plant and harvest crops, other vehicles, such as trains and automobiles, effectively reduced the amount of healthy exercise people partook in each day.

Also, many professions that required large amounts of physical exertion outdoors were replaced by indoor office work, which is often sedentary. Such sedentary behaviors also occur away from work, as television programs and other forms of passive entertainment came to dominate leisure time.

Added to this is the fact that many people eat food that has been processed with salt and sugar to help with its preservation, lower its cooking time, and increase its sweetness. Together, these lifestyle trends have led to increases in lifestyle-related diseases associated with obesity, such as heart disease, diabetes, and certain forms of cancer.

Undergirding the development of modern Europe between the 1780s and 1849 was an unprecedented economic transformation that embraced the first stages of the great Industrial Revolution and a still more general expansion of commercial activity. Articulate Europeans were initially more impressed by the screaming political news generated by the French Revolution and ensuing Napoleonic Wars, but in retrospect the economic upheaval, which related in any event to political and diplomatic trends, has proved more fundamental.

Major economic change was spurred by Western Europe's tremendous population growth during the late 18th century, extending well into the 19th century itself. Between 1750 and 1800, the populations of major countries increased between 50 and 100 percent, chiefly as a result of the use of new food crops (such as the potato) and a temporary decline in epidemic disease.

Population growth of this magnitude compelled change. Peasant and artisanal children found their paths to inheritance blocked by sheer numbers and thus had to seek new forms of paying labor. Families of businessmen and landlords also had to innovate to take care of unexpectedly large surviving broods. These pressures occurred in a society already attuned to market transactions, possessed of an active merchant class, and blessed with considerable capital and access to overseas markets as a result of existing dominance in world trade.

Heightened commercialization showed in a number of areas. Vigorous peasants increased their landholdings, often at the expense of their less fortunate neighbors, who swelled the growing ranks of the near-propertyless. These peasants, in turn, produced food for sale in growing urban markets. Domestic manufacturing soared, as hundreds of thousands of rural producers worked full- or part-time to make thread and cloth, nails and tools under the sponsorship of urban merchants.

Craft work in the cities began to shift toward production for distant markets, which encouraged artisan-owners to treat their journeymen less as fellow workers and more as wage laborers. Europe's social structure changed toward a basic division, both rural and urban, between owners and no owners. Production expanded, leading by the end of the 18th century to a first wave of consumerism as rural wage earners began to purchase new kinds of commercially produced clothing, while urban middle-class families began to indulge in new tastes, such as uplifting books and educational toys for children.

In this context an outright industrial revolution took shape, led by Britain, which retained leadership in industrialization well past the middle of the 19th century. In 1840, British steam engines were generating 620,000 horsepower out of a European total of 860,000. Nevertheless, though delayed by the chaos of the French Revolution and Napoleonic Wars, many western European nations soon followed suit; thus, by 1860 British steam-generated horsepower made up less than half the European total, with France, Germany, and Belgium gaining ground rapidly.

Governments and private entrepreneurs worked hard to imitate British technologies after 1820, by which time an intense industrial revolution was taking shape in many parts of Western Europe, particularly in coal-rich regions such as Belgium, northern France, and the Ruhr area of Germany. German pig iron production, a mere 40,000 tons in 1825, soared to 150,000 tons a decade later and reached 250,000 tons by the early 1850s. French coal and iron output doubled in the same span—huge changes in national capacities and the material bases of life.

Technological change soon spilled over from manufacturing into other areas. Increased production heightened demands on the transportation system to move raw materials and finished products. Massive road and canal building programs were one response, but steam engines also were directly applied as a result of inventions in Britain and the United States. Steam shipping plied major waterways soon after 1800 and by the 1840s spread to oceanic transport.

Railroad systems, first developed to haul coal from mines, were developed for intercity transport during the 1820s; the first commercial line opened between Liverpool and Manchester in 1830. During the 1830s local rail networks fanned out in most western European countries and national systems were planned in the following decade, to be completed by about 1870. In communication, the invention of the telegraph allowed faster exchange of news and commercial information than ever before.

New organization of business and labor was intimately linked to the new technologies. Workers in the industrialized sectors labored in factories rather than in scattered shops or homes. Steam and water power required a concentration of labor close to the power source. Concentration of labor also allowed new discipline and specialization, which increased productivity.

The new machinery was expensive, and businessmen setting up even modest factories had to accumulate substantial capital through partnerships, loans from banks, or joint-stock ventures. While relatively small firms still predominated, and managerial bureaucracies were limited save in a few heavy industrial giants, a tendency toward expansion of the business unit was already noteworthy.

Commerce was affected in similar ways, for new forms had to be devised to dispose of growing levels of production. Small shops replaced itinerant peddlers in villages and small towns. In Paris, the department store, introduced in the 1830s, ushered in an age of big business in the trading sector.

Urbanization was a vital result of growing commercialization and new industrial technology. Factory centers such as Manchester grew from villages into cities of hundreds of thousands in a few short decades. The percentage of the total population located in cities expanded steadily, and big cities tended to displace more scattered centers in Western Europe's urban map.

Rapid city growth produced new hardships, for housing stock and sanitary facilities could not keep pace, though innovation responded, if slowly. Gas lighting improved street conditions in the better neighborhoods from the 1830s onward, and sanitary reformers pressed for underground sewage systems at about this time. For the better-off, rapid suburban growth allowed some escape from the worst urban miseries.

Rural life changed less dramatically. A full-scale technological revolution in the countryside occurred only after the 1850s. Nevertheless, factory-made tools spread widely even before this time, as scythes replaced sickles for harvesting, allowing a substantial improvement in productivity.

Larger estates, particularly in commercially minded Britain, began to introduce newer equipment, such as seed drills for planting. Crop rotation, involving the use of nitrogen-fixing plants, displaced the age-old practice of leaving some land fallow, while better seeds and livestock and, from the 1830s, chemical fertilizers improved yields as well. Rising agricultural production and market specialization were central to the growth of cities and factories.

The speed of Western Europe's Industrial Revolution should not be exaggerated. By 1850 in Britain, far and away the leader still, only half the total population lived in cities, and there were as many urban craft producers as there were factory hands. Relatively traditional economic sectors, in other words, did not disappear and even expanded in response to new needs for housing construction or food production. Nevertheless, the new economic sectors grew most rapidly, and even other branches displayed important new features as part of the general process of commercialization.

Geographic disparities complicate the picture as well. Belgium and, from the 1840s, many of the German states were well launched on an industrial revolution that brought them steadily closer to British levels. France, poorer in coal, concentrated somewhat more on increasing production in craft sectors, converting furniture making, for example, from an artistic endeavor to standardized output in advance of outright factory forms. Scandinavia and the Netherlands joined the industrial parade seriously only after 1850.

Southern and eastern Europe, while importing a few model factories and setting up some local rail lines, generally operated in a different economic orbit. City growth and technological change were both modest until much later in the 19th century, save in pockets of northern Italy and northern Spain. In eastern areas, western Europe's industrialization had its greatest impact in encouraging growing conversion to market agriculture, as Russia, Poland, and Hungary responded to grain import needs, particularly in the British Isles. As in eastern Prussia, the temptation was to impose new obligations on peasant serfs laboring on large estates, increasing the work requirements in order to meet export possibilities without fundamental technical change and without challenging the hold of the landlord class.

The Industrial Revolution was a major turning point in history which was marked by a shift in the world from an agrarian and handicraft economy to one dominated by industry and machine manufacturing. It brought about a greater volume and variety of factory-produced goods and raised the standard of living for many people, particularly for the middle and upper classes. However, life for the poor and working classes continued to be filled with challenges. Wages for those who labored in factories were low and working conditions could be dangerous and monotonous. Children were part of the labor force.

They often worked long hours and were used for such highly hazardous tasks as cleaning the machinery. Industrialization also meant that some craftspeople were replaced by machines. Additionally, urban, industrialized areas were unable to keep pace with the flow of arriving workers from the countryside, resulting in inadequate, overcrowded housing and polluted, unsanitary living conditions in which disease was rampant. The conditions for the working-class gradually improved as governments instituted various labor reforms and workers gained the right to form trade unions.

The factory system was a child of the Industrial Revolution and developed and advanced during its course in the 18th and 19th century. It replaced the cottage industry which was more autonomous with individual workers using hand tools and simple machinery to fabricate goods in their own homes. The invention of the water powered frame by Richard Arkwright in the 1760s led to the formation of the first factories along the rivers in Britain.

In 1771, Arkwright built his first factory at Cromford. He built many small cottages close to it to employ labour from far and across, preferring weavers with large families so that women and, especially their children, could work in the factory. By 1779, he had over 800 people with timed jobs, shifts and factory rules. The factory system generated a fortune for its few owners and his template caught like wild fire. Improvements in the steam engine and power loom further incentivized cheaper energy and better machines; and this positive loop fed the Industrial Revolution.

Capitalism refers to an economic system based upon private ownership of the means of production and their operation for profit. With political control over the colonies and rise in technological innovations, Capitalism was on a rise in Britain. Factory owners and others who controlled the means of production rapidly became very rich and had more money to invest in technology and more industry. In those times only the wealthy could vote in Britain with about 3 percent allowed to vote.

Industrial capitalists gradually replaced agrarian land owners as leaders of the nation's economy and power structure. With economic and political power they were in many ways the new rulers of the nation. Great Britain, in which the Industrial Revolution originated, was followed by other nations including Belgium, France, Germany and the United States. Soon the capitalists became the leaders in numerous countries across the world.

The rise of cities was one of the defining and most lasting features of the Industrial Revolution. In pre-industrial societies almost 80% of people lived in rural areas dependent on farming and animal husbandry. The growth in population due to the agriculture revolution and the rise in industry had reduced the opportunities in the rural areas causing large migrations to the industrialized cities. The population of Britain almost doubled in the 18th century.

By the end of the century 1 in 10 Britons lived in London which had a population of 1 million. In 1771, Manchester had a population of only 22,000. Over the next fifty years, its population exploded and reached 180,000. By 1850, more people were living in cities than in villages. The number of cities with populations of more than 20,000 in England and Wales rose from 12 in 1800 to nearly 200 at the close of the century. This trend was seen all around as other parts of the world industrialized.

For many skilled workers, the quality of life decreased a great deal in the first 60 years of the Industrial Revolution. Skilled weavers, for example, lived well in pre-industrial society as a kind of middle class. They tended their own gardens, worked on textiles in their homes or small shops, and raised farm animals. They were their own bosses. The Industrial Revolution was the shift of primarily agrarian societies to industrialized societies.

The contrast was stark especially for the first few generations of factory workers who knew of life in the country as compared to life in the industrial cities. With almost no laws for the new age and power centered with the wealthy; the new working class in the factories suffered. Their neighborhoods were bleak, crowded, dirty and polluted. The condition of hand skilled workers deteriorated and there was little or no scope to supplement their income with gardening or communal harvesting. During the first 60 years there was little scope for recreation.

Many slums were formed, there was extensive child labour and many people were lost to disease and hazardous working conditions. In 1849, 10,000 people died of cholera in three months in London alone. Tuberculosis claimed 60,000 to 70,000 lives in each decade of the 19th century. In the first 60 years, the situation in general was bleak for many as may be seen even today in developing countries.

Historians disagree about the increase in wages of the working class in the first phase of the Industrial Revolution but there is general agreement that, adjusted for inflation, the wages stayed steady from 1790 to 1840. A rise of about 50 percent is observed between 1830 and 1875 in Britain. There was a very gradual rise of the middle class in the cities, mostly towards the end of the 19th century. The society had always been divided in two classes: the aristocrats born into their lives of wealth; and low-income commoners born in the working classes.

The new urban industrial towns slowly created a plethora of new jobs such as big shopkeepers, bank clerks, insurance agents, merchants, accountants, managers, doctors, lawyers and teachers. Purchasing power increased and total national income multiplied 10 times in Britain in 100 years by the end of the 19th century. As the wealth shifted in the hands of the businessmen, there was more opportunity for enterprising, shrewd and brilliant ideas. There were also many rags to riches stories which inspired people to work harder.

The rise in materialism and consumerism was one of the primary fallouts of the Industrial Revolution. Money, be it gold, paper or plastic is a mode of exchange and it derives its value from the goods and services someone is willing to offer for it. With the rise of Industry more goods were being produced leading to the development of the nation.

At the same time, competitive hand skilled industries were slowly wiped out due to political and economic reasons. As production kept on increasing over decades and centuries, it required a proportional increase in demand. Feeding on the basic human desire to have more, a cycle of more consumption and more production was fired up which led to rise of materialism and consumerism.

Industrial Revolution itself was primarily driven by the rise in technology which forever changed the face of the world leading us into the modern era. The external combustion steam engine powered railways, factories and inspired the internal combustion engine and the automotive industry. Energy demands led to electricity and electric based appliances. Telegraph led to the telephone to finally the internet and mobile technology. There are numerous examples to suggest the giant strides humanity took in the field of technology during and as a consequence of the Industrial Revolution.

The government majorly favoured the wealthy in the early part of the Industrial Revolution. Even children were not spared and in the early 1860s, an estimated one-fifth of the workers in Britain's textile industry were younger than 15. With a large population that felt exploited under a few wealthy capitalists, social tensions gradually increased. The condition of the working class became such a cause of concern that it led to the rise of socialism. Socialism is a theory which advocates that all people are equal and should have shared ownership of the country's wealth. The most influential socialist thinker was undoubtedly an economist and philosopher named Karl Marx (1818-1883). Marx spent most of his time in England understanding and critiquing the established capitalist system of those times.

His ideas challenged the very foundations of the capitalist world, inspiring many uprisings against the model. Marxism and Communism as economic models are however widely rejected in the world today due to their lack of success wherever implemented.

India and China had been the dominant economies of the world for centuries. In the beginning of the 18th century they accounted for close to 50 percent of the world GDP. By the 18th century the British, Dutch, Portuguese and French were involved in sea trade with India for over a century and were now aware and involved to some extent in the politics of the region. With victory in the battles of Plassey and Buxar in mid-18th century the British gained considerable power in India outplaying its rivals. The drain of wealth from India gained momentum with these victories through laws, taxes and de-industrialization among many other things.

In the 17th and 18th centuries demand for Chinese goods (particularly silk, porcelain, and tea) in Europe created a trade imbalance between Qing Imperial China and Great Britain. Opium had been a problem for China and it had already been illegal to smoke and sell opium in China since 1729. The British with control on India, auctioned opium in Calcutta to licensed merchants, who shipped the opium to British-owned warehouses in the free trade area in Canton (Guangzhou), China. From there, the opium was smuggled by Chinese traders to the rest of the country, often with the help of corrupt customs officers outside the British zone.

The influx of drugs drained the Chinese economy and impaired its population. This led to the Opium wars in 1839 – 1842 and 1856 – 1860 which Qing China lost to Britain. These victories allowed Britain to force opium into the Chinese markets in return for Chinese goods. Thus opium trade was made more open leading to further decline of the nation.

Pollution and environmental damage were the obvious consequences of the industrialized world and the consumerism that followed it. The rise of the machines required vast amounts of energy to fuel them, and fossil fuels like coal and petroleum were burned to energize the industry resulting in smog and air pollution.

Chemicals were necessary for various processes leading to the fast rise in the development of industrial parks based on the chemical manufacturing of such items as dyes, plastics and pharmaceuticals. Cities were densely populated and forests and farmlands were cleared to make room for railroads and other infrastructure. Waste was dumped in rivers and cities were highly polluted.

The Great Stink in London in August 1858 was a noted event during which hot weather exacerbated the smell of untreated human waste and industrial effluent that was present on the banks of River Thames. The continued advancement of technology allowed large corporations to dictate the industrial landscape, and to have a far-reaching adverse effect on the environment.

2. THE EMERGENCE OF THE FOURTH INDUSTRIAL REVOLUTION

The fourth industrial revolution is the current and developing environment in which disruptive technologies and trends such as the Internet of Things (IoT), robotics, virtual reality (VR) and artificial intelligence (AI) are changing the way we live and work.

The third industrial revolution, sometimes called the digital revolution, involved the development of computers and IT (information technology) since the middle of the 20th century. The fourth industrial revolution is growing out of the third but is considered a new era rather than a continuation because of the explosiveness of its development and the disruptiveness of its technologies. According to Professor Klaus Schwab, Founder and Executive Chairman of the World Economic Forum and author of The Fourth Industrial Revolution, the new age is differentiated by the speed of technological breakthroughs, the pervasiveness of scope and the tremendous impact of new systems.

The Fourth Industrial Revolution heralds a series of social, political, cultural, and economic upheavals that will unfold over the 21st century. Building on the widespread availability of digital technologies that were the result of the Third Industrial, or Digital, Revolution, the Fourth Industrial Revolution will be driven largely by the convergence of digital, biological, and physical innovations.

Like the First Industrial Revolution's steam-powered factories, the Second Industrial Revolution's application of science to mass production and manufacturing, and the Third Industrial Revolution's start into digitization, the Fourth Industrial Revolution's technologies, such as artificial intelligence, genome editing, augmented reality, robotics, and 3-D printing, are rapidly changing the way humans create, exchange, and distribute value.

As occurred in the previous revolutions, this will profoundly transform institutions, industries, and individuals. More importantly, this revolution will be guided by the choices that people make today: the world in 50 to 100 years from now will owe a lot of its character to how we think about, invest in, and deploy these powerful new technologies.

It's important to appreciate that the Fourth Industrial Revolution involves a systemic change across many sectors and aspects of human life: the crosscutting impacts of emerging technologies are even more important than the exciting capabilities they represent. Our ability to edit the building blocks of life has recently been massively expanded by low-cost gene sequencing and techniques such as CRISPR; artificial intelligence is augmenting processes and skill in every industry; neurotechnology is making unprecedented strides in how we can use and influence the brain as the last frontier of human biology; automation is disrupting century-old transport and manufacturing paradigms; and technologies such as block chain and smart materials are redefining and blurring the boundary between the digital and physical worlds.

The result of all this is societal transformation at a global scale. By affecting the incentives, rules, and norms of economic life, it transforms how we communicate, learn, entertain ourselves, and relate to one another and how we understand ourselves as human beings.

Furthermore, the sense that new technologies are being developed and implemented at an increasingly rapid pace has an impact on human identities, communities, and political structures. As a result, our responsibilities to one another, our opportunities for self-realization, and our ability to positively impact the world are intricately tied to and shaped by how we engage with the technologies of the Fourth Industrial Revolution.

This revolution is not just happening to us—we are not its victims—but rather we have the opportunity and even responsibility to give it structure and purpose.

As economists Erik Brynjolfsson and Andrew McAfee have pointed out, this revolution could yield greater inequality, particularly in its potential to disrupt labor markets. As automation substitutes for labor across the entire economy, the net displacement of workers by machines might exacerbate the gap between returns to capital and returns to labor. On the other hand, it is also possible that the displacement of workers by technology will, in aggregate, result in a net increase in safe and rewarding jobs.

All previous industrial revolutions have had both positive and negative impacts on different stakeholders. Nations have become wealthier, and technologies have helped pull entire societies out of poverty, but the inability to fairly distribute the resulting benefits or anticipate externalities has resulted in global challenges.

By recognizing the risks, whether cybersecurity threats, misinformation on a massive scale through digital media, potential unemployment, or increasing social and income inequality, we can take the steps to align common human values with our technological progress and ensure that the

Fourth Industrial Revolution benefits human beings first and foremost.

We cannot foresee at this point which scenario is likely to emerge from this new revolution. However, I am convinced of one thing—that in the future, talent, more than capital, will represent the critical factor of production.

With these fundamental transformations underway today, we have the opportunity to proactively shape the Fourth Industrial Revolution to be both inclusive and human-centered. This revolution is about much more than technology—it is an opportunity to unite global communities, to build sustainable economies, to adapt and modernize governance models, to reduce material and social inequalities, and to commit to values-based leadership of emerging technologies.

The Fourth Industrial Revolution is therefore not a prediction of the future but a call to action. It is a vision for developing, diffusing, and governing technologies in ways that foster a more empowering, collaborative, and sustainable foundation for social and economic development, built around shared values of the common good, human dignity, and intergenerational stewardship. Realizing this vision will be the core challenge and great responsibility of the next 50 years.

Humanity continues to embark on a period of unparalleled technological advancement. The next 5, 10 and 20 years will present both significant challenges and opportunities. Private sectors, governments, academics and entrepreneurs are all seeking the roadmap for navigating these profound changes in the world of work. Such a road map must be created collaboratively by all stakeholders.

At its core, an industrial revolution can be characterized by advancements in technology that humanity applies to improve the process of production. But in reality, it means so much more.

The first three industrial revolutions brought to the world water and steam power, electricity and digitization. With every industrial revolution comes refining shifts to social, economic, environmental and political systems that truly alter the course of humanity. Some of these shifts are foreseen, and others are completely unforeseen .

Today, a fourth industrial revolution unfolds. The Fourth Industrial Revolution is bringing technologies that blur the lines between the physical, digital and biological spheres across all sectors. Technologies like artificial intelligence (AI), nanotechnology, quantum computing, synthetic biology and robotics will all drastically supersede any digital progress made in the past 60 years and create realities that we previously thought to be unthinkable. Such profound realities will disrupt and change the business model of each and every industry.

One of the most immediate and impactful outcomes of technological evolution is the vast advancement in automation. Every day, more manual process become automated, and as technology continues to accelerate, so will automation.

As a result, the world of work and labor market demand are rapidly changing. According to McKinsey, up to 375 million workers may need to change their occupational category by 2030, and digital work could contribute $2.7 trillion to global GDP by 2025. Faced with the scale of the unstoppable shifts in workforce demands, we must address the challenges associated with workforce transformation, starting by taking an in-depth look at its impact on the world of work. Four key impact areas should be considered:

For most global industries (e.g., logistics, financial, manufacturing, aerospace, etc.), advancements in AI, robotics, 3D printing and the internet of things will put a great deal of pressure on companies to automate in order to remain competitive in a global landscape. This will require companies to have a solid understanding of the way these technologies impact their industries and how they can ensure organizational agility to adapt to these changes. Increased global competitiveness will accelerate cost pressure, which will lead to substantial downsizing or reassignment of a large contingent of workers. McKinsey estimates that up to 800 million individuals may be displaced by automation by 2030.

There are four factors of production that fuel economic growth: land, labor, capital and enterprise. Today, the world is attaining only 52% of its entrepreneurial capacity, and this number is declining year over year. Large, established enterprises have a significant advantage in the future of work than smaller companies due to their ability to adapt to technological changes. However, this is not a recipe for long-term, sustainable economic success. The world must focus on supporting independent entrepreneurs, as small and midsize businesses are the fuel of most economies of the world today.

Technology will continue to change societal values. Today, more than 36% of the U.S. workforce are freelancers for reasons including autonomy, flexibility and extra income. Co-working spaces are exploding in popularity and are often fully subscribed before opening their doors. Technology has enabled people to work anytime, anywhere. By 2027, more than half of American workers will be freelancing.

Education and training: Part-and-parcel with economic development is one's ability to access training for employment. Naturally, tectonic shifts are happening in the education space. Students are less interested in stale curriculums and keener to take shorter, skills-based training that is more relevant to today's workplace. Employers are focusing on the skills required to achieve their business objectives and remain competitive and agile, which requires them to ensure their employees the necessary training to fill these skills gaps. Workers, naturally, need to acquire skills "on demand" to adapt to their changing roles and responsibilities.

Despite the challenges we face, we also possess an unprecedented possibility to apply an abundance mindset to solving the challenges. The Fourth Industrial Revolution will provide us with an opportunity to learn and teach new skills, build new jobs requiring unique skills combinations that don't exist today, explore talent that we didn't know about and, in doing so, grow our businesses and create a new generation of workers that are highly skilled in more diverse areas. The question is, how do we get there?

Collaborations among the private sector, academia and policymakers will be essential to navigate the future of work as we go through these profound moments. Schools need to work with businesses and the public sector to develop on-demand, relevant, adaptable curriculums and focus on teaching skills; governments need to utilize advanced technologies to generate real-time and predictive insights on the labor market in order to develop sound policies, programming and budgets; companies need to hire for competencies over credentials and, more importantly, take the lead in supporting existing workforces' upskilling and lifelong learning.

The Fourth Industrial Revolution is changing how we live, work, and communicate. It's reshaping government, education, healthcare, and commerce—almost every aspect of life. In the future, it can also change the things we value and the way we value them. It can change our relationships, our opportunities, and our identities as it changes the physical and virtual worlds we inhabit and even, in some cases, our bodies.

Education and access to information can improve the lives of billions of people. Through increasingly powerful computing devices and networks, digital services, and mobile devices, this can become a reality for people around the world, including those in underdeveloped countries.

The social media revolution embodied by Facebook, Twitter, and Tencent has given everyone a voice and a way to communicate instantly across the planet. Today, more than 30% of the people in the world use social media services to communicate and stay on top of world events.

These innovations can create a true global village, bringing billions more people into the global economy. They can bring access to products and services to entirely new markets. They can give people opportunities to learn and earn in new ways, and they can give people new identities as they see potential for themselves that wasn't previously available.

"The Fourth Industrial Revolution, finally, will change not only what we do but also who we are. It will affect our identity and all the issues associated with it: our sense of privacy, our notions of ownership, our consumption patterns, the time we devote to work and leisure, and how we develop our careers, cultivate our skills, meet people, and nurture relationships." —Klaus Schwab, The Fourth Industrial Revolution

Online shopping and delivery services—including by drone—are already redefining convenience and the retail experience. The ease of delivery can transform communities, even in remote places, and jumpstart the economies of small or rural areas.

In the physical realm, advances in biomedical sciences can lead to healthier lives and longer life spans. They can lead to innovations in neuroscience, like connecting the human brain to computers to enhance intelligence or experience a simulated world. Imagine all that robot power with human problem-solving skills.

Advances in automotive safety through Fourth Industrial Revolution technologies can reduce road fatalities and insurance costs, and carbon emissions. Autonomous vehicles can reshape the living spaces of cities, architecture, and roads themselves, and free up space for more social and human-centered spaces.

Digital technology can liberate workers from automatable tasks, freeing them to concentrate on addressing more complex business issues and giving them more autonomy. It can also provide workers with radically new tools and insights to design more creative solutions to previously insurmountable problems.

However, while the Fourth Industrial Revolution has the power to change the world positively, we have to be aware that the technologies can have negative results if we don't think about how they can change us.

We build what we value. This means we need to remember our values as we're building with these new technologies. For example, if we value money over family time, we can build technologies that help us make money at the expense of family time. In turn, these technologies can create incentives that make it harder to change that underlying value.

People have a deep relationship with technologies. They are how we create our world, and we have to develop them with care. More than ever, it's important that we begin right. We have to win this race between the growing power of the technology, and the growing wisdom with which we manage it. We don't want to learn from mistakes.

Biotechnology can lead to controversial advances such as designer babies, gene drives (changing the inherited traits of an entire species), or implants required to become competitive candidates for schools or jobs. Innovations in robotics and automation can lead to lost jobs, or at least jobs that are very different and value different skills.

Artificial intelligence, robotics, bioengineering, programming tools, and other technologies can all be used to create and deploy weapons.

Social media can erase borders and bring people together, but it also can also intensify the social divide. And it gives voice to cyber-bullying, hate speech, and spreading false stories. We have to decide what kind of social media rules we want to create, but we also have to accept that social media is reshaping what we value and how we create and deploy those rules.

In addition, being always connected can turn into a liability, with no respite from the continuous overload of data and connections.

Artificial intelligence is unleashing a whole new level of productivity and augmenting our lives in many ways. As in past industrial revolutions, it can also be a disruptive force, dislocating people from jobs and surfacing questions about the relationship between humans and machines.

It's inevitable that jobs are going to be impacted as artificial intelligence automates a variety of tasks. However, just as the Internet did 20 years ago, the artificial intelligence revolution is going to transform many jobs—and spawn new kinds of jobs that drive economic growth. Workers can spend more time on creative, collaborative, and complex problem-solving tasks that machine automation isn't well suited to handle.

However, workers with less education and fewer skills are at a disadvantage as the Fourth Industrial Revolution progresses. Businesses and governments need to adapt to the changing nature of work by focusing on training people for the jobs of tomorrow. Talent development, lifelong learning, and career reinvention are going to be critical to the future workforce.

People are asking whether the Fourth Industrial Revolution is the road to a better future for all. The power of technology is increasing rapidly and facilitating extraordinary levels of innovation. And as we know, more people and things in the world are becoming connected. But that doesn't necessarily pave the way for a more open, diverse, and inclusive global society.

The lessons of previous industrial revolutions include the realization that technology and its wealth generation can serve the interests of small, powerful groups above the rest. Powerful new technologies built on global digital networks can be used to keep societies under undue surveillance while making us vulnerable to physical and cyberattacks. These are the challenges we can face to make sure the combination of technology and politics together don't create disparities that hinder people.

According to the World Economic Forum Global Risks Report 2017, "the Fourth Industrial Revolution has the potential to raise income levels and improve the quality of life for all people. But today, the economic benefits of the Fourth Industrial Revolution are becoming more concentrated among a small group. This increasing inequality can lead to political polarization, social fragmentation, and lack of trust in institutions. To address these challenges, leaders in the public and private sectors need to have a deeper commitment to more inclusive development and equitable growth that lifts up all people".

Many people around the world haven't yet benefited from previous industrial revolutions. As the authors of Shaping the Fourth Industrial Revolution point out, at least 600 million people live on smallholder farms without access to any mechanization, living lives largely untouched by the first industrial revolution. Around one-third of the world's population (2.4 billion) lack clean drinking water and safe sanitation, around one-sixth (1.2 billion) have no electricity—both systems developed in the second industrial revolution. And while the digital revolution means that more than 3 billion people now have access to the Internet, that still leaves more than 4 billion out of a core aspect of the third industrial revolution.

The means that as we appreciate and engage with the exciting technologies of the Fourth Industrial Revolution, we must work to ensure that the opportunities they bring are well-distributed around the world and across our communities. In particular, we must help those who missed out on the huge increases in quality of life that the first, second, and third industrial revolutions provided.

"Let us together shape a future that works for all by putting people first, empowering them and constantly reminding ourselves that all of these new technologies are first and foremost tools made by people for people." — Klaus Schwab, The Fourth Industrial Revolution.

We value the ability to control what is known about us, and yet we are living in a world where tracking every individual's personal information is key to delivering more intelligent, personalized services. For example:

Facebook tracks what you do so that it knows which content and advertisements are most relevant to you.

Smartphones track your location, and you can share that information with apps that recommend places to eat or shop. Retailers analyze your purchase history to recommend products and offer coupons to stimulate more sales.

In the future, you'll walk into a store and the salesperson will immediately have your name, credit rating, marital status, and past purchases flashed to their augmented-reality virtual screen.

Technological advances are also broadening the scope of surveillance. In the UK today, an estimated 6 million CCTV cameras are recording activity all over the country. Advances in computing power and artificial intelligence can potentially enable law enforcement agencies to track suspected terrorists by analyzing social networks, government records, and other data.

In the future, billions of 3D-printed "smart dust" cameras floating in the air can monitor the activities of humans. From traffic reports to natural disasters, such technology can keep us safer. But it also can watch us when we do not want to be watched.

For consumers, businesses that are transparent about their data collection practices and that prioritize consumer privacy can win our loyalty.

Public trust in business, government, the media, and even technology is falling. This is a crisis that is dividing societies and creating instability around the world.

The technologies of the Fourth Industrial Revolution themselves are neutral, but are they being applied in ways that build trust? Are consumers going to trust that new artificial intelligence and robotic systems can make their lives better, or are they going to be fearful of the machines and those who control them? Are citizens going to trust the institutions and service providers who collect and maintain their data?

For the Fourth Industrial Revolution to generate trust, everyone contributing to it (including you) must collaborate and feel a connection to common objectives. More transparency into how we govern and manage this technology is key, as are security models that boost our confidence that these systems won't be hacked, run amok, or become tools of oppression by those who control them.

The innovations in artificial intelligence, biotechnology, robotics, and other emerging technologies are going to redefine what it means to be human and how we engage with one another and the planet. Our capabilities, our identities, and our potential will all evolve along with the technologies we create.

In the coming decades, we must establish guardrails that keep the advances of the Fourth Industrial Revolution on a track to benefit all of humanity. We must recognize and manage the potential negative impacts they can have, especially in the areas of equality, employment, privacy, and trust. We have to consciously build positive values into the technologies we create, think about how they are to be used, and design them with ethical application in mind and in support of collaborative ways of preserving what's important to us.

This effort requires all stakeholders—governments, policymakers, international organizations, regulators, business organizations, academia, and civil society—to work together to steer the powerful emerging technologies in ways that limit risk and create a world that aligns with common goals for the future.

You, as a person, citizen, employee, investor, and social influencer, are a critical stakeholder in the Fourth Industrial Revolution. Sharing your thoughts on the technologies and what you value as this revolution unfolds is essential. The world we create through technologies can shape our lives and is the one we pass on to the next generation.

"The Fourth Industrial Revolution can compromise humanity's traditional sources of meaning—work, community, family, and identity—or it can lift humanity into a new collective and moral consciousness based on a sense of shared destiny. The choice is ours." —Klaus Schwab, The Fourth Industrial Revolution.

3. BENEFITS AND RISKS OF THE FOURTH INDUSTRIAL REVOLUTION

Around the world, the ground appears to be shifting beneath our feet as we hurtle through a period of unnerving change, marked by disturbing weather patterns, xenophobic protectionism, mass migration, failed states, science denial, cyberterrorism, a loss of faith in institutions and many other factors.

What's more, global energy markets are in flux. Supply chains are dauntingly complicated. And in most industrialized nations, societies are ageing, which raises tough questions about who will pay for the extended retirement of so many people.

At the same time, the power of technology, with its abundant promise, has accelerated due to dramatic gains in data storage, processing power and algorithm-driven analytics. Month after month, new systems, applications and business models surface and then explode into the market, offering radical new solutions in domains such as health and transport, even while disrupting long-established businesses and throwing countless people out of work.

In a fraught period that has been dubbed the Fourth Industrial Revolution, governments, civil society and the private sector have a duty to ensure that nations such as Canada are prepared for this new world and its dizzying challenges. Amid so much flux, one point is certain: we can either catch this competitive wave or be swamped by it.

According to the World Economic Forum's Founder and Executive Chairman, Professor Klaus Schwab, the first three industrial revolutions set the stage for the fourth: the early 19th century era of rail, mechanization and steam; the electricity and mass production revolution in the late 19th and early 20th centuries; and the emergence of semiconductors, computers and networks since the 1960s. The exponential acceleration of computing technology that has marked this phase is inflicting massive change on long-established industries, professions and institutions, including the structures of government.

The technologies driving these changes - big data, machine learning, blockchain, the Internet of Things, advanced materials, quantum computing and 3D printing - are complex, esoteric and profoundly disruptive. To those of us operating in sectors being remade by these innovations, the pace can feel as if we're perched on slippery stones in the middle of a rapidly moving river, looking for a way to cross.

The Fourth Industrial Revolution is changing how we grow, buy and choose what we eat.

Shift will happen everywhere, from routine shop floor work to the tasks performed by professionals such as doctors, lawyers, and accountants. Large companies will continue to find themselves toppled by fleet-footed start-ups that begin with little capital and few hard assets but plenty of technical savvy.

Historically, such periods of technology-driven upheaval have brought productivity gains, investment, growth, improvements in quality of life, and increases in longevity and health. There's no reason to believe that the Fourth Industrial Revolution, like the three that preceded it, will fail to deliver these same long-term benefits, especially in a world where billions of people still don't have electricity.

Yet some of these technologies, especially those that automate routine tasks, may trigger job losses. That future is around the corner, in fact. A recent McKinsey & Company study predicts that almost half the time workers spend on their jobs can already be replaced with existing technologies.

The transition period we're entering is extremely fluid - and frankly, scary. Everyone needs that next paycheck, a safe place to sleep, money for groceries. If Fourth Industrial Revolution technologies yield chronic unemployment, will political and social unrest follow? We only have to look at disenfranchisement of un- or underemployed American blue-collar workers to understand why we're living in a period of rising nationalism, xenophobia, and protectionism.

The alarming job-loss scenarios have also prompted warnings and calls for corrective policy. Tesla founder Elon Musk wants governments and civil society actors to ensure that machine learning systems are deployed ethically. Microsoft founder Bill Gates wants governments to tax robots to compensate for mass worker displacement.

After a decade of flat productivity, the arrival of the Fourth Industrial Revolution (4IR) is expected to create up to $3.7 trillion in value to global manufacturing. A few years back, experts noted that the changes associated with the 4IR would come at an unprecedented rate yielding incredible results for those who truly embraced them.

Still, the hockey stick of benefits has not kicked in yet—while all companies are making efforts to adopt technology, most of the production industry (~70%)

remains in pilot purgatory (where technology pilots last for extended periods of time, and companies do not take the final step of scaling up viable technologies). Less than 30% of manufacturing companies are actively rolling out Fourth Industrial Revolution technologies at scale.

The World Economic Forum, in collaboration with McKinsey, has undertaken a global search and assessment for "4IR production lighthouses"—sites representing the most advanced sub-section of the companies who are actively deploying 4IR technologies at scale. Leveraging these lighthouse sites, we will create an inclusive learning platform with aims on becoming the next "go-to" system for advanced manufacturing and bring the world closer to capturing the lion's share of the benefits.

Most manufacturing lines still look a lot the same way they did 10, 20 or even 30 years ago. Operators clock in, have a brief conversation with their crew and shift supervisor, and then operate a machine or tool for 8-12 hours before heading home. Depending on the day, the machine may break and need maintenance or an adjustment, the line settings may need to be modified for a specific product or run, or the operator may need to step away to resupply the line or be trained on a new procedure.

The fourth industrial revolution (4IR) – the convergence and interpenetration of digital technologies, bio, nano, info and things - promises great benefits, from advances in mobile money and energy, to homes and healthcare, for example combatting cancer. Large scale deployment of digital, AI and automation should give us big gains in productivity, which should mean more prosperity, a bigger cake for everyone to share, as well as many social gains .

These are just some of the reasons why we are now seeing feverish excitement among investors, entrepreneurs and governments, with AI, and the 4IR, centre stage in industrial strategies from China to the UK.

Unfortunately this isn't a straightforward or straightforwardly good revolution. As many observers have pointed out, the 4IR risks widening the divide between vanguards and the rest; accelerating job destruction ahead of job creation; and introducing potentially serious threats to personal privacy and cybersecurity.

The first industrial revolution probably did more to benefit humanity than any other event in history (certainly as measured by its effects on life expectancy, income, and freedom).

But in its first decades it also did huge harm, driving millions into poverty, ill-health and vulnerability to crime in cities like Manchester and Chicago. Only when complementary innovations (like sewers), social innovations (like welfare states) and institutional innovations (like democracy) came along, were the benefits widely spread .

All of the key technology fields that are now proving so exciting have been primarily driven by military or military related investment: machine learning; computer vision; robotics; and what we now call the Internet of Things (the one partial exception is natural language processing).

Meanwhile, although a good proportion of the new applications in business provide useful advances in mobility and efficiency, too many are more intriguing than useful. Examples like refrigerators that warn you when you need to buy more milk are emblematic of a technological revolution that risks being diverted into purposes that are either trivial or harmful.

Work with the World Economic Forum (through the Global Future Council on innovation and entrepreneurship) has suggested four fundamental shifts that are now overdue to make this a more useful revolution:

The first is a shift in ends and purposes. There's an urgent need to redirect investment in the technologies of the 4IR more towards the most important human needs, including healthcare, mobility and education, rather than warfare and advertising. Making labour markets work well; helping refugees integrate into new societies; or reducing crime. These are all promising areas for investment that have so far had only small crumbs of funding by comparison with other fields like optimizing recommendation engines or guiding missiles. Nesta's investments in companies using AI for education or jobs are good examples. Making this shift will be good for society; but it will also mean fewer failed investments for business too .

The second is a shift in means and participation. The 4IR is largely being shaped by small groups of people in big companies, a few governments and universities. The rest of the population is observers. We need, instead, to open up the 4IR to millions of entrepreneurs, innovators, makers, and citizens, and use new tools to make it easier for them to shape this revolution. There are many examples of how this is being done well, from Nesta's Longitude Explorer Prize (which backs 11-16 year olds with Internet of Things innovations) to the hundreds of maker spaces around the world .

The third is a shift of ethos, to humanize the 4IR. The 4IR technologies don't only risk making many people redundant, literally, they also risk amplifying the worst sides of human nature - as has happened already with social media, which, at times, reinforces tendencies towards aggression, addiction and compulsive behaviour. We need to multiply applications that do the opposite and reinforce our dispositions to cure, care and relate. We need different ethics and aesthetics for technology to make them more engaging, and more emotionally intelligent in ways that are reciprocal not manipulative.

The fourth is a shift to take seriously the need for complementary innovations. Some of these will emerge in the field of regulation (like the many forms of anticipatory regulation we are developing); some will be social innovations (like new approaches to data of the kind being experimented with by Decode); and some are institutional innovations (like the Machine Intelligence Commission we have proposed).

There are many good examples of initiatives that embody the different spirit described here – in fields as diverse as farming and mental health, finance and care. But these remain, generally, small scale.

If these shifts don't happen people will understandably come to see the 4IR as a threat. That happened to many technologies in the past, from nuclear power to genetically modified crops. Their advocates assumed that the world would be grateful for technological breakthroughs. But too often they failed to ask basic questions about who stood to benefit, and who faced risks.

4IR offers an explosion of tools for intelligence. In cultivating them we shouldn't suspend our own intelligence, and forget to ask the questions that matter most.

The Fourth Industrial Revolution is reshaping every sphere of human life — from government to commerce; from education to healthcare. It is even impacting human values, opportunities, relationships and identities by modifying virtual as well as physical worlds of human beings.

- **POSITIVE IMPACTS :-**
1. For the advancements in latest technological innovations, the power and types of digital devices, computing devices and networks are rapidly

developing day by day. This is making education and various information easily accessible.

2. Gradual evolution of technologies and scientific innovations are leading to the creation of new educational disciplines, which is, finally leading to more scopes for better opportunities. Fourth Industrial Revolution is enhancing the facilities for the development as well as innovation of new skills. Fourth Industrial Revolution emphasizes growth of knowledge and thirsts for learning. Application oriented courses are more preferable than bookish education.

3. Due to the continuous technological development, online social media, such as, Twitter, LinkedIn, Facebook etc. are becoming more and more active. Everyone is able to express and highlight their views about any contemporary incident or event easily before the world through these social media platforms.

4. Communication is becoming easier steadily. Through WhatsApp, IMO, Messenger etc., people can comfortably connect with and contact their relatives, friends or anyone across the world. Video

calling or chat is helping in compressing distances and making people happier.

5. Lands are not essential for building markets. Online shopping sites and quick delivery services are making commodities accessible at home as well as increasing economic benefits. The online customer service agents also provide smart recommendations to customers.

6. The World is becoming a global village, where billions of people as well as products are easily accessible.

7. Progress in medical sciences, neurosciences etc., due to Fourth Industrial Revolution, are leading to healthier lives; advanced intellectual and mental capability; and longer life spans.

8. Agriculture is also influenced by the Fourth Industrial Revolution. Greater amounts of Crops can be yielded with the help of Bioengineering. With the help of the machines, powered by artificial intelligence, measuring crop populations and detection of weeds or plant pests are also becoming easier. Robotic sprayers are also available for the application of herbicides.

9. Due to the advancement in digital technology, workers are becoming free from automatable jobs and can engage themselves for solving complicated business issues. This is making them more autonomous.
10. Carbon emissions, road fatalities and insurance costs minimize because of the advancements in automotive safety due to the advancing technologies of Fourth Industrial Revolution.
11. No need of standing on road and waiting for transportations. People can book cars or vehicles online and avail them at their doorsteps. Due to the grace of Fourth Industrial Revolution, autonomous or driverless vehicles may be available soon.
12. With the help of online banking facility, people do not need to go to banks for transactions or other important works at bank. Maximum bank-works may be accomplished from home.
13. E-Governance is also possible in the era of Fourth Industrial Revolution with the help of new technological innovations. The new technology also helps the government in modernizing executive organizations and functions. E-Governance ensures accountability and transparency as well as

strengthens the relationship between the government and the citizens.

14. Online jobs provide people to work from and earn at home.
15. Fourth Industrial Revolution emphasizes self-employment.

- **NEGATIVE IMPACTS :-**

1. Over-reliance on technology is decreasing the will of human beings for using own intellect and physical power.
2. Social media is increasing distances between a person and his family members as well as the physical society. Virtual world is becoming more preferable than the physical world and this is creating social divide.
3. Social media is not always helpful because it is a medium for spreading news, among which some or many may be false; the false news create annoyances.
4. Privacy of an individual is not at all totally private in this era of Fourth Industrial Revolution due to the grace of technological advancements in tracking system. Every activity of a human being can be traced through digital devices, like CCTV Cameras,

smart phones etc. Social media platforms, such as, Facebook, Twitter etc. as well as online shopping sites, such as, Flipkart, Amazon etc. collect every information from name and Date of Birth to Credit Card or bank details of an individual before creating a profile or account.

5. Cyber bullying and hate speech are other negative impacts of social media which are gradually increasing in the era of Fourth Industrial Revolution. Cyber-attacks are also not unlikely due to the gradual progress in internet facility. It should not be forgotten that hacking is not always ethical; hacking may also harm our overall security.

6. Overuse of data and connections is overloading the network services.

7. People are least interested to go to the market, to jog under the open sky or to visit someone's home, among others because the technological evolution enable them to shop at online shopping sites, to jog on treadmills and to contact with people through social media respectively, among others. Children are more interested in mobile games than outdoor games due to the grace of smart technologies. This is affecting human health badly, physical as well as

mental, because the movement of the human body and intake of fresh air is decreasing. Excessive uses of smart phones and digital games are causing hindrance in physical and mental growth of children.

8. The scopes of employment are at stake because of the advancements in automotive and robotic technology. Human skills are becoming invaluable in front of artificial intelligence. Machines are more favored than human beings.

9. Concentration of wealth among small group of people is constantly increasing. This is creating inequality among people economically as well as socially. Thus social fragmentation, political polarization and lack of trust in institutions are inevitable.

10. Competitive environment of Fourth Industrial Revolution, sometimes, causes emotional frustration as well as affect mental balance. This may lead to suicidal tendency, anxiety, insomnia and other neurological diseases.

11. Bioengineering, artificial intelligence, programming tools, robotics etc. may also be used for destructive purposes.

12. Controversial innovations, due to biotechnology, like gene drives or implants to increase the efficiency of a human being, as well as designer babies, among others cannot be ignored.
13. Climate change is another negative impact of Fourth Industrial Revolution. Massive industrialization, rapid development in technologies, increased urbanization, excessive deforestation, limitless resource depletion, desertification, rapid population growth, water scarcity, food insecurity etc. are hampering the natural environmental balance of earth.
14. Rapid migration of people for better opportunity and clustering at one place, leaving another isolated creates pressure at one part of the earth while the development of other part lags behind. This causes environmental imbalance as well as unequal global development, which causes social inequality globally. Growth in inequality may invite conflicts, social tensions and violent extremism.

4. THE IMPACT OF FOURTH INDUSTRIAL ROVUTION ON LABOR

Many commentators are increasingly talking about the potential impact of the "4 th Industrial Revolution." It will change how we live and how we work, how the economy works and how we are governed. For example, a Citi and Oxford University joint report in 2016 estimated that 57% of jobs across the OECD are at risk of automation, the Financial Times reported in 2016 1 that between 2000 and 2010, of all the jobs lost in the US, over 85% were lost to new technologies, and the Bank of England estimated that two thirds of all jobs are capable of being automated within 20 years.

Regardless of the specific data, an indisputable fact is that the 4th Industrial Revolution has already come, and the current workforce is already feeling the heat. Whilst the business world is already discussing and preparing for how this revolution will affect their businesses, dubbing it "Industry 4.0", the wider societal impacts of this new revolution has not, to date, been discussed in depth nor planned for. Past Industrial Revolutions have forced society to undergo major and often painful processes of adaptation, for example from rural, largely agricultural societies, to urban, industrial societies, and then to post-industrial societies dealing with the loss of traditional industries and sources of employment.

The societal impacts of the 4th Industrial Revolution also appear likely to be far-reaching, resulting not only in the social and economic impacts of the loss of many current jobs, but also fundamental, and increasingly volatile shifts in the nature of work and future jobs, and in how public and private services will be delivered.

The Third Sector has always been at the forefront of meeting societal challenges and needs, ranging from mental health and wider health and social care services, to services for older people, to meeting transport needs in less accessible areas which private or public sector operators could not provide on a commercial or cost-effective basis. The Third Sector is therefore well-placed to play a key role in meeting the wide range of changing societal needs likely to emerge from the 4th Industrial Revolution.

Building on the first Industrial Revolution which used water and steam power to mechanize production, the second which used electric power to create mass production and the third, which used electronics and information technology to automate production; the 4th Industrial Revolution is taking automation to new levels, blurring the lines between the physical, digital, and biological spheres and using technologies to perform tasks previously carried out by humans, ranging from piloting vehicles to 'rules-based' jobs in areas such as accounting and law.

When we compare it with previous Industrial Revolutions, we find the dramatic differences between the fourth Industrial Revolution and the other three. In its scale, scope, and complexity, the transformation will be unlike anything humankind has experienced before.

The 4th Industrial Revolution is not merely a prolongation of the Third Industrial Revolution but rather a new and distinct revolution. Firstly, people can continuously produce new information and generate new knowledge in the mining of information. The possibilities of billions of people connected by mobile devices, with unprecedented processing power, storage capacity, and access to knowledge, are unlimited.

We can record a person's daily life through their mobile phone location. When this data is monitored for a long period of time, we can get to know a person's lifestyle habits, such as their work place, the supermarkets they shop in, the restaurants where they dine, the times they do so, and even their personal preferences. This technology will allow the intelligence level of machines to increase through continuous data accumulation and analysis.

Secondly, the Industrial Revolution represents not only a huge advance in technology and in the improvement of productivity, but will also transform modes of production and the relationships between elements of production processes. The 4th Industrial Revolution, by enabling the complete communication of all relevant information at every stage in the production chain, creates separate production sectors for each process and informs how they relate to each other, bringing together such processes as inventory taking, improving production efficiency, saving energy and reducing emissions, thus making the manufacturing industry part of the information industry.

At the same time, it can make production flexible and allow mass customization, enabling different products to be produced in a production line, which will revolutionize the warehousing, transportation and the whole manufacturing industry. Thirdly, the 4th Industrial Revolution will spawn a new economic form, the 'sharing economy.'

A typical example of sharing economy is ride-hailing online services, such as Uber and the Chinese Didi service, which allow customers to obtain taxis services from private car owners. The impact of this new form is disruptive, not only to the taxi industry, but also the whole transportation industry. (Maybe in the near future, we won't need drivers at all and unpiloted vehicles will fill the streets.)

The impacts of the sharing economy are not limited to online ride-hailing services, but also include the shared space service, e.g. Airbnb, and the global online work platform, e.g. AAwork. From the shared motors and houses, to the shared umbrellas, basketballs, toys, clothing and jewelry, the sharing economy is constantly updating, and will be very profound and revolutionary. Last but not least, as the economists Erik Brynjolfsson and Andrew McAfee have pointed out, the revolution could yield greater inequality, particularly in its potential to disrupt labor markets.

With the growth of automation, robots and computers will replace workers across a vast spectrum of industries. Low-skill/low-pay jobs will disappear and the poor will face tougher challenges, which in turn will lead to an increase in social tensions. In a strict sense, this is not a unique feature of the fourth Industrial Revolution. Historically, Industrial Revolutions have always begun with greater inequality followed by periods of political and institutional change. However, mankind will face a more serious challenge in this revolution, because it is robots and computers that take our jobs, not the flow of labour between different sectors.

The characteristics of the fourth Industrial Revolution are destined to bring about different impacts on employment, which are no longer confined to one industry, but all industries. At the same time, a lot of jobs will disappear, but there will be a lot of new job requirements.

It is expected that more than 65% of children entering primary school today will end up working in completely new jobs that currently do not exist when they enter the workplace 15 years from now. As the changes brought by the social media, digital publications and e-commerce, the most in-demand occupations did not exist 10 or even five years ago.

According to the Future of Employment report, around 47 percent of total US employment is in the high risk category. People may be more concerned about what types of jobs are at high risk than specific Numbers. So which jobs are at greatest risk? What jobs will be safe in the future? Researchers at Oxford University published a widely referenced study in 2013 on the likelihood of computerization for different occupations. Out of around 700 occupations, here are the top 30 most risky occupations having a 98-99 per cent chance of being automated in the future:

1 Telemarketers
2 Title Examiners, Abstractors, and Searchers
3 Sewers, Hand
4 Mathematical Technicians
5 Insurance Underwriters
6 Watch Repairers
7 Cargo and Freight Agents
8 Tax Preparers
9 Photographic Process Workers and Processing Machine Operators
10 New Accounts Clerks
11 Library Technicians 12 Data Entry Keyers
13 Timing Device Assemblers and Adjusters
14 Insurance Claims and Policy Processing Clerks
15 Brokerage Clerks
16 Order Clerks
17 Loan Officers
18 Insurance Appraisers, Auto Damage
19 Umpires, Referees, and Other Sports Officials
20 Tellers
21 Etchers and Engravers

22 Packaging and Filling Machine Operators and Tenders

23 Procurement Clerks

24 Shipping, Receiving, and Traffic Clerks

25 Milling and Planning Machine Setters, Operators, and Tenders, Metal and Plastic

26 Credit Analysts

27 Parts Salespersons

28 Claims Adjusters, Examiners, and Investigators

29 Driver/Sales Workers

30 Radio Operators

On the other hand, the following list comprises the top 30 most safe occupations with a 0.66 percent or less probability of being computerized based on current technology.

1 Recreational Therapist

2 First-Line Supervisors of Mechanics, Installers, and Repairers

3 Emergency Management Directors

4 Mental Health and Substance Abuse Social Workers

5 Audiologists

6 Occupational Therapists

7 Orthotists and Prosthetists

8 Healthcare Social Workers

9 Oral and Maxillofacial Surgeons

10 First-Line Supervisors of Fire Fighting and Prevention Workers

11 Dietitians and Nutritionists

12 Lodging Managers

13 Choreographers

14 Sales Engineers

15 Physicians and Surgeons

16 First-Line Supervisors of Transportation and Material-Moving Machine and vehicle operators

17 Instructional Coordinators

18 Psychologists, All Other

19 First-Line Supervisors of Police and Detectives

20 Dentists, General

21 Elementary School Teachers, Except Special Education

22 Medical Scientists, Except Epidemiologists

23 Education Administrators, Elementary and Secondary School

24 Podiatrists

25 Clinical, Counseling, and School Psychologists

26 Mental Health Counselors

27 Fabric and Apparel Patternmakers

28 Set and Exhibit Designers

29 Human Resources Managers

30 Recreation Workers

The 4th Industrial Revolution is creating a demand for new jobs while eliminating some of the jobs highlighted in the above reports. In the short term, mankind will face a great challenge and the jobless will soar. As the research of Erik Brynjolfsson and Andrew McAfee from the MIT Sloan School of Management, starting around 2011, technology has fueled productivity but not fueled job growth—quite the opposite, actually. McAfee and Brynjolfsson indicate that part of the reason is that our skills aren't keeping up with technological advances.

However, people should not be too pessimistic because, in the long run, the 4th Industrial Revolution will create more wealth and additional jobs elsewhere in the economy and the number of new jobs will grow dramatically. Due to the difficulties for some people to adapt to the new job requirements and master new job skills, the real problem that people are facing is structural unemployment, not lack of job opportunities.

Firstly, the jobs that are most at risk are those which "are on some level routine, repetitive and predictable", as

Martin Ford, futurist and author of Rise of the Robots: Technology and the Threat of a Jobless Future explains, because they are possible to replicate through Machine Learning algorithms.

Richard Johnston from Ulster University's Economic Policy Centre said: "Sectors like manufacturing, logistics and retail and wholesale and some of the lower skilled occupations within are the most vulnerable to being replaced by some technology or machinery or robots." For any business owner, the pursuit of profit maximization is the most important goal and reducing cost is an important factor to consider.

No matter how low people's wages are they'll never be able to compete with the robots and machines, no salary, no break and no illness. Machines are better at the job: The National Institute of Standards predicts that "machine learning can improve production capacity by up to 20%" and reduce raw materials waste by 4%. In fact, many highly routine occupations are being replaced today. Telemarketing, for example, which ranks the first according to The Future of Employment report, has a 99% probability of automation.

The vast majority of people have received irritating robocalls. Library technicians, whose responsibilities are to compile records, sort and shelve books, remove or repair damaged books, register patrons, and check materials in and out of the circulation process, also have a 99% probability of automation. These things can be solved by existing technologies, like Amazon's fulfillment centers where people work with carefully coordinated robotic machines. Previously, Amazon workers walked around shelves looking for products, but now robotic shelves rearrange themselves to bring products to the worker.

These sorting techniques will also be used for book sorting, which will save a lot of manpower and provide additional efficiencies. Secondly, jobs that were once regarded as secure jobs, such as office workers, administrative personnel, and even law, will be hit hardest. Future technological advances will enable people to work more flexibly allowing people to work at home or in the office, and during working hours or off-duty hours.

Employers, for their part, are also happy to choose this kind of flexible work pattern to reduce office expenses. Each employee may not need a fixed working seat, but can share working space with others. This can save thousands of pounds in furniture, office equipment and supplies, and utilities savings compared to similar workers who have fixed places in the office. Moreover, The "Gig Economy", a labour market characterized by the prevalence of short-term contracts or freelance work as opposed to permanent jobs, may become more and more common as part of the evolution of job flexibility.

Some call this trend the "Uberization" of work, as the Future of Jobs report mentioned. Remote platforms, on which freelance or independent workers sell work to a customer, such as Upwork, Freelancer, TaskRabbit, Clarity, and 99designs facilitate the Gig Economy. In the future, people may have several jobs for a number of companies simultaneously rather than working for one big corporation and normal full-time work seems to be in trouble.

Thirdly, jobs based on big data analysis, such as credit analysts, financial advisers, mathematical technicians, will face huge risks. The total amount of digital data in circulation was estimated to be 4.4 zettabytes in 2013, while this number is predicted to increase tenfold to 44 zettabytes in 2020 and grow faster and faster in the future.

Computers have a distinct advantage over the human brain in dealing with big data. Computers can store, access, analyze, interpret and draw meaningful inferences from big data with more accuracy and efficiency than the human brain. The floating point arithmetic ability of the most ordinary computer can compute more than 10 billion times a second, which is far more than the computing power of the human brain.

The best chess-trained computers can strategize many moves ahead, problem-solving far more deftly than can the best chess-playing humans. It is no wonder that Google's AI beat the Go world champion. Computers enjoy other advantages over people. Computers have better memories, so they can be fed a large amount of information, and can tap into all of it almost instantaneously. For computers, the word "oblivion" has no meaning, but for humans, no one can remember everything that happens to them. In China, a variety show, 'Stand to the end', is very popular.

The best player, who beats all the other people taking part must then complete with an intelligent robot to answer ten questions. Up to now, after more than 5 episodes, no one can win this robot, because it never makes a mistake. Because of the nature of computers, jobs based on the analysis of big data can be done better by computers than of humans. Our credit ratings are based on recording all of our personal behaviors and analyzing these by computer. There is no need to meet a financial adviser face to face because we can just get the computer's advice.

Firstly, jobs requiring a level of human interaction or guiding robot behavior will be very popular in the future, such as first-line supervisors of mechanics, installers, and repairers and first-line supervisors of transportation and material-moving machine and vehicle operators. These skilled people possess in-demand skills on how to run and manage new technologies like robotics, autonomous transport, new energy supplies and 3D printers.

People will be working with robots and machines, not competing with them. In the future, there may be machine trainers who teach machines to work better. Preparing the necessary data sets in advance is essential for any artificial intelligence.

For instance, in medicine, they may teach robots and machines how to detect diseases with existing marker cases as demonstrations of what to look for. Since the trend towards the development of artificial intelligence appears inevitable, artificial intelligence developers are likely to become very popular. At present, there is a huge gap in demand for artificial intelligence engineers, which directly leads to a $345,000 annual salary for advanced AI researchers in DeepMind Technologies Limited. AI testers will test intelligent robots, spotting problems and errors, and correcting error codes especially in the early stages.

The second area is occupations that involve building complex relationships with people, especially customer-facing jobs to supply personalized services, such as sales engineers, mental health and substance abuse social workers and mental health counselors. These jobs need high interpersonal skills, teamwork and leadership, which computers cannot go beyond, such as dealing with coordination of people and communication, and divergent communications.

It does not help to have a robot give people a pep talk. The clergy only has a 0.81% probability of automation, according to data from The Future of Jobs. Aging populations will drive a dramatic increase in spending on healthcare and other personal services. By 2030, there will be at least 300 million more people aged 65 years and older than there were in 2014. This will create significant new demand for a range of occupations, including doctors, nurses, and health technicians but also home-health aides, personal-care aides, and nursing assistants. Moreover, Health care will go from general to personal. Doctors are already using computers and other high-tech devices to improve health care. As data becomes more readily available, extensive and personalized, it will revolutionize the way doctors diagnose disease and treat patients. The online data will help doctors have access to more information on patients and link patients' wellness to their lifestyle which leads to personal services. To some extent, the treatments have to be done in thousands of ways because everyone is unique. Thirdly, jobs that are highly unpredictable would be very difficult to be replaced by robots and machines, such as emergency management directors and repairers.

They are technically difficult to automate because machines are good at repetitive tasks. The other reason is that they often command relatively lower wages, which makes automation a less attractive business proposition. Employment of emergency management directors is projected to grow 6 percent from 2014 to 2024, about as fast as the average for all occupations, as the forecast of Bureau of Labor Statistics.

The last area is occupations on training and education. Upcoming workforce transitions could be very large. According to the Future of Employment report, around 47 percent of total US employment will disappear in the future. Meanwhile, new jobs will be available but people need to find their way into these jobs. The changes in net occupational growth or decline imply that a very large number of people may need to shift occupational categories and learn new skills in the years ahead. Many people may have to re-train several times during their working life. Work that requires a high degree of imagination, creative analysis, and strategic thinking is harder to automate.

Creativity will determine whether human beings can develop sustainably in the future, and creativity should be nurtured by education. Josefino Rivera, Jr., educator indicated education will be not just taking in information and sharing it back, but also figuring out what to do with that information in the real world.

The 4th Industrial Revolution is significantly different from the previous Industrial Revolutions. It will completely change everyone and every aspect of life. As Stanford University academic Jerry Kaplan writes in Humans Need Not Apply: today, automation is "blind to the color of your collar."

It doesn't matter whether you're a factory worker, a financial advisor or a professional flute-player: automation is coming for you. This applies not just to individuals but even more so to organizations and the implications for the third sector in the coming age of change is profound. The 'third sector', belonging neither to the public sector nor to the private sector, covers a range of different organizations with different structures and purposes, which often is described as the charity and voluntary sector, non-governmental organisation, non-profit organizations, community sector, civil society sector and so on.

This sector as a whole has evolved in scope and scale in the last hundred years. According to the 2017 report conducted by the Johns Hopkins Center on Civil Society Studies, the global civil society sector today has mushroomed into a global workforce of 350 million professionals and volunteers, "outdistancing major industries in the scale of its workforce and in its contribution to social and economic life."

Put differently, if the global civil society workforce were a country, it would be the third most populated country in the world following China and India. Moreover, every Industrial Revolution brings a lot of social changes and social problems. Typically most third sector organizations devote themselves either to a particular social issue which needs solving or to a particular group in society who requires support and representation.

Many charities focus on issues surrounding social services, housing, education, human rights, community development, international development, health and medicine, and conservation and environment. From this point of view, the impact of the 4th Industrial Revolution on the third sector is definitely profound. Is the third sector equipped to successfully navigate the challenges and opportunities of the 4th Industrial Revolution? Can they ensure the 4th Industrial Revolution advances in a manner that maximizes benefits and minimizes harms to people and the society? The bar chart below shows the main areas of work conducted by voluntary, community and social enterprise organizations -weighted ranked scores.

It may be surprising that the main areas of work that the third sector is currently engaged in are highly coincident with the work required for the 4 th Industrial Revolution. For example, occupations that involve building complex relationships with people, such as healthcare and other personal services will increase dramatically in demand.

Charities always play a significant part in health and wellbeing area, providing expert healthcare, conducting research, raising awareness, supporting patients, and promoting mental health and well-being. Health is the third largest charity sub-sector by expenditure, with 6,626 health charities spending £4bn in 2011/12. In the UK much health care is provided in the public sector by the National Health Service (NHS).

However, there is also a significant provision of supplementary care by third sector bodies. Charities, compared with the public sector, are able to think more holistically, taking into account physical, emotional and environmental challenges and tackling the root causes of health inequality.

Charities are at the heart of the communities they support: some directly deliver health and social care services; many work with a range of beneficiaries to provide care around daily problems. It even has the potential to allow a greater shift in focus onto the causes rather than the symptoms of problems, lessening the onset of preventable diseases and potential demand for treatment in future.

These services cater to the needs of personalized medical services in the future. A similar situation exists in the work areas of disability and older people. At the dawn of the 4th Industrial Revolution, more and more people are aware of the challenges of job losses and unemployment. Many people may have to re-train several times during their working life, which will lead to a growing demand for education and training.

At present, much education is provided by the public sector. However, some independent schools and colleges are third sector bodies and some forms of education, such as private tuition, is provided on a for-profit basis. The third sector has a distinct advantage over statutory agencies in personalized education and training. Some third sector organizations are more flexible than statutory agencies in the services they provide. For example, colleges only have intakes twice per year. This causes problems for the service users who may not be able to progress to college provision when they are ready to do so. Any "thirst for learning" that users have engendered can be lost while individuals wait until the next intake.

Moreover, often working with individuals or small groups, third sector organizations are also able to tailor their learning offer in a way that public services, who generally provide for a much wider range of needs and abilities, find much more challenging. Through working with specific groups they are also able to develop in-depth knowledge of specific needs and expertise in designing services to meet those needs. In this perspective the third sector will meet the requirements of changing labor market better in the future.

The third sector has a tremendous opportunity in the future. At the same time, the 4th Industrial Revolution will transform the third sector significantly.

The advent of the 4th Industrial Revolution, will bring new societal problems and issues and societal problem and issues are often the focus of the work of the third sector. So, people might see more new charities and NGOs popping up to tackle social issues. As Rob Acker, CEO of Salesforce.org, predicts social good organizations can scale like never before because we're more connected than ever before.

Historically, a lack of resources and funding has plagued the social sector, for example, one in five smaller charities is struggling to survive, but technology can help small organizations make a big impact. The cloud is helping to break down barriers to entry. With increased access to data, populations that were previously unreachable can now be tapped and connected with particular causes without having to drastically increase overhead costs.

First, AI offers the potential for a more personalized service by freeing up community sector workers to do more people-oriented roles and a lot less of the grunt work. They will speak to residents, talking to customers, help the elderly, do all those compassionate, emotionally needy jobs that machines can't and shouldn't do. Second, services will become more timely, and even predictable. There's a huge opportunity for nonprofits to reach more people than ever before and connect with their donors, volunteers, students and constituents in real-time from anywhere. Nonprofits can instantly reach their community of donors and volunteers to help with urgent matters that may mean the difference between life and death. For example, a nonprofit focused on the humanitarian crisis, could identify the specific location and number of refugees coming into different countries, and preemptively send the appropriate level of aid and supplies.

According to the report, the State of Charities and Social Enterprises 2015, generating income and achieving financial sustainability is the most pressing challenge facing charity chief executives, and that two fifths of chief executives in large charities are concerned about a reduction in public or government funding. However, with the 4IR this problem will be solved thoroughly. Organizations can also start to organize and understand these communities better than ever before, resulting in deeper engagement.

According to the recently released Connected Nonprofit Report, 65% of donors would give more money if they felt their nonprofits knew their personal preferences—and 75% of volunteers would give more time. With deeper engagement, these organizations will start to see increases in donations and volunteer time, which directly impacts their mission. The 4th Industrial Revolution is going to redefine what it means to be human and how we engage with one another and the planet. The third sector can use technology to find and connect with more people who need their services, understand their communities on a deeper level, and supply better services to achieve more goals.

At the same time, their capabilities and potential will all evolve along with the technologies. In the coming decades, the third sector will face both increased opportunities and challenges (e.g. an aging population) . They must recognize and seek to address future societal challenges,, particularly in the areas of social inequality and unemployment. This effort requires all stakeholders—governments, policymakers, international organizations, regulators, business organizations, academia, and civil society—to work together to steer the powerful emerging technologies in ways that limit risk and create a world that aligns with common goals for the future.

5. THE IMPACT OF FOURTH INDUSTRIAL REVOLUTION ON HIFHER EDUCATION

Higher education in the fourth industrial revolution (HE 4.0) is a complex, dialectical and exciting opportunity which can potentially transform society for the better. The fourth industrial revolution is powered by artificial intelligence and it will transform the workplace from tasks based characteristics to the human centered characteristics.

Because of the convergence of man and machine, it will reduce the subject distance between humanities and social science as well as science and technology. This will necessarily require much more interdisciplinary teaching, research and innovation.

Today, all graduates face a world transformed by technology, in which the Internet, cloud computing, and social media create different opportunities and challenges for formal education systems. As students consider life after graduation, universities are facing questions about their own destiny especially employment. These technologies powered by artificial intelligence are so much transforming the world that social concepts such as "post-work" are more and more defining the present period. This period requires certain skills that are not exactly the same as the skills that were required in the third industrial revolution where information technology was the key driver.

These skills are critical thinking, people management, emotional intelligence, judgment, negotiation, cognitive flexibility, as well as knowledge production and management. Our starting point is to investigate the three current megatrends as well as their consequences.

We argue that one insightful lens of today's life is based on intelligent technology that is powered by artificial intelligence. Fast changes in physical (e.g., intelligent robots, autonomous drones, driverless cars, 3D printing, and smart sensors), digital (e.g., the internet of things, services, data and even people) and biological (e.g., synthetic biology, individual genetic make-up, and bio-printing) technologies, and generally in the way we work, we learn, and we live, make it a crucial force for economic competitiveness and social development.

The Fourth Industrial Revolution With the waves of above mentioned breakthroughs in various domains, we gradually find ourselves in the midst of the fourth industrial revolution which is driven by artificial intelligence (AI) and cyber physical systems (CPS) (Marwala, 2007). To understand the first industrial revolution was catalyzed by

Newton when he formulated his laws of motion. Because from then onwards motion was better understood and quantified, it was possible to design stem engines that mechanized much of the work that was traditionally done by humans. The second industrial revolution was catalyzed by Faraday and Maxwell who unified magnetic and electric forces and this led to electricity generation and electric motor which were instrumental in the assembly lines that have come to dominate many industries.

The third industrial revolution was catalyzed by the discovery of a transistor which ushered the electronic age that gave us computers and internet. The fourth industrial revolution will revolutionize industries so substantially that much of the work that exists today will not exists in 50 years (Marwala et al., 2006). The next subsections describe hallmarks that characterize the fourth industrial revolution.

The 4th industrial revolution digitizes and vertically integrates processes across the entire organisation. It also integrates horizontally all the internal processes from suppliers to customers. Put simply, it epitomizes a shift in paradigm shift from 'centralized' to 'decentralized' production, whereby machines no longer simple 'process' the product, but they are seamlessly integrated into the information network, the business partners and customers.

In other words, the idea of consistent digitization and linking of all productive units in an economy is emphasized in the 4th industrial revolution age.

The main goal of mass higher education was targeting transfer of skills and preparation for a wide variety of technical and economic roles. Post-Massification Higher education has gradually progressed from the elite phase to mass higher education and then to post-massification stages. Many advanced and some developing economies enjoy the tertiary participation rates of over 50%. Another characteristic of this trend is internationalization of both students and staff. According to a report from OECD, with demographic changes, international student mobility is expected to reach 8 million students per year by 2025.

Currently, adapting population to rapid social and technological change remains the main goal of many countries. The core mission of higher education remains the same whatever the era. The goal of higher education is to ensure quality of learning via teaching, to enable the students to get the latest knowledge through exploratory research, and to sustain the development of societies by means of service.

One of the principal tasks of every university is to educate the youth. Therefore, it is necessary to implement appropriate teaching strategies and to organize work in a way that fosters learning. This has implications on adaptable learning programmers, better learning experience, and lifelong learning attitude.

The journey towards global competition in the higher education requires institutions to put a huge amount of effort into research and development (R&D). Experts believe these forces range from new technology deployment to global cooperation and collaboration.

To sustain the competitive position among world higher education system, we need to radically improve educational services. In particular, we need to drive much greater innovation and competition into education.

Higher Education in the Fourth Industrial Age (HE 4.0) With its speed and breadth, the question brought by the aforementioned megatrends and the subsequent fourth industrial revolution for higher education it is important for nations to understand the impact of these changes to all areas of our lives including higher education.

The plurality of wearable devices produced indicates an early sign of another technology. Education establishments have to act now to realize wearables' huge potential to revolutionize the way we teach and train students and how they learn as well. Take numerical simulation, it is a very useful tool for engineers to analyze and predict the condition of real-world physical systems. In the era of the 4th industrial revolution, when the existence of cyber-physical systems become a new norm, numerical simulations play an ever-increasing important role in both education and practical applications. Within the realm of numerical simulation, finite element analysis (FEA) is a versatile technique which has been practiced in various engineering fields such as analyzing buildings (Marwala et al., 2017; Marwala, 2012; Marwala, 2010).

Modern FEA is often accomplished with the assistance of computers. As a result, students can understand key concepts more intuitively, and engineers can conduct complex modeling and interpret results easily. Nevertheless, such setup has limited the FEA processes in an entirely virtual and offline environment.

These limitations in turn deprive the human perception of many physical characteristics (e.g., scale, context, spatial qualities, and materials). With the advancement of some wearable technologies, say augmented reality (AR), a user's sense and interaction with the physical world can be enhanced thereby creating a virtual laboratory.

AR can supplement reality via superimposing computer-generated information over the physical context in real time which can facilitate results exploration and interpretation.

Teaching has long been constrained by the following scenario: students needed to gather in a lecture hall to hear the professor or sit around a table to discuss with peer fellows. Technology innovation is relaxing those constraints, however, and brining radical change to higher education. Massive open online courses, or MOOCs, is a form of education that provides stand-alone instruction online (Xing, 2015). Though much experimentation lies ahead, MOOCs threaten different universities in distinct ways.

Two big factors underpin a university's costs: physical proximity requirement and productivity limitation. Because of the need for physical proximity enrolling more students is expensive considering the increase in buildings and instructors. Because of productivity limitation, the maximum number of students that can be compressed into lecture venues and exam-marking rosters are limited. MOOCs can eliminate these obstacles by working completely differently: off campus and online model; and once an online course is created, teaching extra students become an advantage.

lack innovative talent, especially at the high end. To fully grasp the opportunity of another wave of industrialization, a country's higher education system should not only focus on training knowledge-based skilled person, but have a good look at cultivating innovative talent, especially high-level scientists and technologists. These scientists must be trained in an interdisciplinary environment where technologists should understand humanities and social science and vice versa.

Microeconomics is an important subject in higher education which has both social and practical value (Marwala, 2013; Marwala, 2014; Marwala, 2015). But most of its concepts exhibit a high level of abstraction which often imposes great difficulties for students to learn it. In many situations, the concepts are isolated, without comprehensive understanding the correlations of each piece of knowledge point on the whole picture.

The aftermath of this learning process is that only parts are recognizable by students, while the comprehension of the overall working mechanism is paralyzed. In this regard, the main objective for a lecturer is to let students acquire the conceptual knowledge (i.e., essential relationship between knowledge fragments and their functions in the whole knowledge system) which is applied to not only microeconomics but many other subjects as well. To address this issue, we believe a generalized blended learning (i.e., mixed e-learning and face-to-face learning methodology) may contribute to this. It is well-known that virtual environments offer great educational value in the process of information transmission and interactive participation, either in real time (e.g., video conferences), or non-simultaneous participants involvement (e.g., forums and chats).

In such process, the face-to-face teaching and evaluation can be used to develop analytical expressions and problem solving capabilities related to mathematical matters. Lecturers at this stage can get physical feedback about the effectiveness of their knowledge transmission to students. Then the understanding of some specific conceptual issues are further assessed and reinforced via online graphic representations and multiple choice test questions and this offers students an advantage of reviewing their results immediately. In closing rather than fighting against these new technologies and the associated novel teaching patterns, higher education systems need to look at how they can accept them and transform the teaching and learning environment to the benefit of both students and academics.

Open innovation, refers to the combination of humans and computers to form distributed systems for the purpose of accomplishing innovative tasks that neither can be done alone. Despite the debate about accuracy, information science has begun to build on some early successes (say, Wikipedia) to demonstrate the potential of evolving open innovation that can model and resolve wicked problems at the junction of economic, environmental, and socio-political systems.

A typical open innovation process includes: (a) Micro-tasking under crowdsourcing mechanism where the respective strengths of a crowd and machines can be magnified. (b) Designated workflows guide crowd-workers to use and augment the information offered by workers at the previous step. (c) To create problem-solving ecosystems, researcher can then combine the cognitive processing of many ordinary contributors with machine-based computing to establish faithful models of the complicated, interdependent systems that underlie the world's most demanding tasks.

Under higher education in the fourth industrial age, a country's higher education system should put innovation, evolutionary and revolutionary, high on its agenda. In general, innovations based on existing technologies are so-called evolutionary type; while revolutionary type of innovations focuses are inventions of new technologies. Ideally, hybrid innovation is a sound strategy but it is difficult to implement. Established academics are often victims of their own accomplishments. Leading scholars have long succeeded by exploring new research domains that could lead to incremental research output growth. Emerging researchers have aggressively followed a similar strategy.

As one research area matures and competition increases severely, the degree of research outputs being published in the form of patents or journals inevitably gets very low. Introducing new research directions means going up against entrenched competition (Xing and Gao, 2014). In the era of the 4th industrial revolution, higher education needs to deepen its technology system reforms by breaking down all barriers to innovation.

One noteworthy obstacle is resource allocation for funding different research projects. For those technology innovations that are important for industrialization, re-industrialization, and neo-industrialization, but are unable to profit in the marketplace in the near-term, financial support from institution and government levels should be made available. However, for applied technologies where commercialization is possible, social capital can play an active role (Xing, 2017). Additionally, several other hindrance should also be dealt with properly: First, with its hybrid innovation strategy, higher education practitioners need to have a global perspective.

The trend of world technology development should be well-perceived and thus appropriate plans need to be made. Each stream of innovation resources, internally, locally, regionally, and globally, should be utilized properly. Second, by having various development strategies and incentive policies across different departments, the connectivity among them should be optimized to avoid potential overlapping. Third, the speed of technology transfer needs to be raised to boost the economic and social development.

Driven Research and Development New technological advancements are often ranked as the most important driving force for R&D. Technology-driven R&D comes in many forms and it can mean employing mobile capabilities to improve data acquisition accuracy; using advanced big-data analytics to spot hidden statistical patterns; harnessing artificial intelligence techniques to retool information search, collection, organization, and knowledge discovery, to name just a few. The bottom line, in all cases, is that the advanced technologies can be leveraged across many domains to continue to deliver impact.

Briefly, advanced technologies can bring benefits to higher education R&D in at least four areas: cost and timeline reduction; operation transformation; R&D process enhancement; and, most significant, research direction innovation via the creation of new ideas and theories. Take the example of additive manufacturing (or 3D printing), this new technology can be used to reduce the cost of producing prototypes, which are generally time consuming and cost inefficient in conventional higher education R&D.

This innovation results in both significant efficiencies and more flexible experimental plans which, in turn, lead to uses of the technology where cost had previously been prohibitive in the laboratory environment. In the fourth industrial age, R&D processes with the help of advanced technologies treat functions such as IT and analytics as 'centers of value' rather than of service or cost; nurture partnership attitude; and, more and more frequently, form an agile style of R&D.

One of the most important and difficult task, is to shift higher education R&D culture from an outdated 'waterfall approach' to idea development. Higher education institutions that make this change will become good at absorbing ideas from all kinds of sources.

Speed enables higher education institutions to be aware of research trends as they emerge and catch up with competition. In comparison to commercial R&D house, higher education institutions' overlylong development times are the most-blamed obstacle to generating positive returns on innovation. In practice, fast movers are much more likely to also be strong innovators as they are also more disruptive.

Brainstorming, conceptualization, model design, theoretical proving, experiment settingup, components procurement, prototyping, test conducting, results analysis, and deliverables submission can be organized into teams that work closely with group leaders to quicken responsiveness to emerging research trends. In closing, as we have indicated herein, the strongest innovators and leading researchers draw on swiftness, well-pruned processes, and the exploitation of advanced technology to explore and capture research opportunities.

Any higher education institution thinking about research in the fourth industrial revolution should first determine where its gaps are vis-à-vis the areas mentioned above and make a plan to address those issues. Once the internal house is in an opposite situation, they can begin to scout around for attractive odds for incremental research capacity growth.

We are all aware of the downfall of Blackberry. The Red Queen effect, the requirement of running faster just to remain in the same place, is one of the most commonly cited causes. In such scenario, one rival that successfully adopts a platform-oriented methodology can compress an entire sector's innovation life cycle.

The outcome of the Red Queen effect is that it becomes harder for a competitor to get ahead of the dominant player. In systems thinking, this phenomenon is often referred to escalation. Nowadays, platform concepts are creating an entirely new competition landscape, one that puts ecosystems in face-to-face wrestling. In Service 4.0, the ongoing transformation to platform based competition is led by many forces: educational activities; ubiquitous computing and Internet of things both within and outside campus and the demanding students in terms of customized learning.

That which served institutions well in scientific disciplines and speciality based markets can become their impediments in platform-based environments. Managing platform-based higher education businesses require a complete different mind-set for strategy. In non-education sector, the likes of Alibaba, eBay, Facebook, WeChat, Google, Baidu and Amazon are actively building their empires around the idea of platform thinking. While platform thinking is not new, what is new is that platformcentric styles are turning into the engines of innovation that are spreading a wide variety of unanticipated sectors such as automobiles, manufacturing, fashion, healthcare, publishing, and many others.

In principle, the platform-based business models emphasize more biologically inspired thinking style rather than a mere organizing logic. In a broader view and analogic metaphor manner, higher education institutions need to reconceive their business ecosystems, re-identify their competitive edges, reshuffle their customer pools, reshape themselves as orchestrators, and rebuild service architecture.

University-as-a-Platform (UaaP) gives the current higher education system an opportunity to steer their bread-and-butter businesses towards platform businesses for a better service performance. Key drivers for a successful UaaP include: a) deliver inter-, multi-, and across disciplinary degrees; b) an appropriate blend of service models (e.g., blended learning, MOOCs, etc.); c) the emergence of Internet of everything; d) the integration of routine education activities into software across a plethora of institution system; e) up-to-date digital infrastructure; and f) enhanced connectivity among all parties residing in higher education value chain.

Typically, in the age of 4th industrial revolution, once every couple of decades, a disruptive new technology arises that essentially changes the blueprint of many sectors. In terms of higher education, the massive proliferation of affordable mobile devices, Internet broadband connectivity and rich education content start a trend of transforming how education is delivered.

Cloud computing, amongst other techniques, creates a new way of educating people that might eventually disrupts the existing higher education systems. With the support of education cloud, government decision makers and business practitioners can answer some key strategic questions comprehensively: deliver education in the quickest, most efficient and best affordability form; develop 21st century students' skills and prepare students for the new job market in the most appropriate way; encourage native innovation with the strongest incentives; and share resources across institutions, districts, regions, or the entire country in the smoothest fashion.

When universities think of embracing EaaS, they often imagine profound advertising campaigns, big promotion budgets, and huge amount of infrastructure investment. Fortunately, EaaS has a healthier respect for the students than academicians have for disruptive ways of delivering education service. At the heart of EaaS is the belief that students' needs should be met effectively.

Therefore, when a higher education institution sets out to attract a potential student as a customer, it needs to create an all-round education experience that is genuinely capable of satisfying the customer's needs, although, this process is not as simple as it may seem. EaaS is not the creation of pseudo differences via a change in logo, location, or making vague promises with empty sounding words.

Furthermore, higher education institutions are accountable to a host of stakeholders such as governments, accrediting agencies, the public and private funding sources, academics, management, support staff, and students. An EaaS orientation that translates into an effective education scheme will achieve these broader concerns.

Nevertheless, many institutions adopt EaaS strategy poorly by giving lip service to various stakeholders.

Education and technology has advanced over the past few decades. Many technology-assisted / enhanced educational practices are no longer as simplistic. In Service 4.0, EaaS as a guideline has to discover newer and more advanced strategies to cope with ever-increasing societal complexity.

With the fast pace of the 4th industrial revolution, forging various kinds of institutional linkages, both domestically and internationally, to offer more versatile degree programmes and professional qualifications becomes a must. Among these schemes, the following types stand out and are worth consideration: First, twinning programmes where a local education provider collaborates with a foreign education provider to develop a connected system allowing course credits to be taken in different locations.

On completion of the twining programme, foreign education provider awards a qualification. Second, franchise programmes are a scenario where foreign education provider authorizes a local education provider to deliver their courses / programmes, and the qualification is awarded by the foreign education provider.

Third, double or joint degree is an arrangement where local and foreign education providers cooperate to offer a programme for a qualification that is awarded jointly or from each of them. Fourth, blended learning where local and foreign education providers deliver programmes to enroll students in various mixed forms, e.g., e-learning, online learning or on-site learning. In closing we trust that improving the quality of service in higher education can bring about a significant change in the society.

4. Conclusions

Though the business of higher education remains unchanged since the times of Aristotle, today students still assemble at a scheduled time and venue to listen to the wisdom of scholars. Given the fourth industrial revolution, a new form of a university is emerging that does teaching, research and service in a different manner. This university is interdisciplinary, has virtual classrooms and laboratories, virtual libraries and virtual teachers. It does, however, not degrade educational experience but augment it.

6. THE IMPACT OF FOURTH INDUSTRIAL REVOLUTION ON HEALTHCARE

We are living in a digital world that is often defined as "the Fourth Industrial Revolution (4IR)". According to the World Economic Forum, 4IR is characterized by "a fusion of technologies that is blurring the lines between the physical, digital, and biological spheres". Healthcare is no exception to this, due to the evolution of physical/occupational therapies and home rehabilitation in the context of 4IR.

Broadly speaking, there are two major benefits that 4IR will bring to the healthcare industry through technologies. One is to prevent the disease itself, and the other is to more efficiently manage the disease once it occurs. Both are directly related to cost savings; digital rehabilitation focuses on the latter by complementing conventional physical and occupational therapies.

When it comes to rehabilitation, home rehabilitation is becoming more important than ever. Many studies suggest that it is one of the most important factors that determine the success of the rehabilitation. "Anytime anywhere" rehabilitation outside the clinic setting has been always sought after, and with the 4IR, home rehabilitation is getting close to achieving that ideal. The following components are the examples showing how it can digitally impact the rehabilitation world.

With the large volume of patient data to be accumulated and analyzed, researchers and clinicians will be able to gather new types of knowledge about an individual's rehabilitation. For example, we will be able to learn about the patient's behavior patterns (i.e. how often, how long, when does the patient work on their rehab) and group them into different categories to develop individualized training schedules at home.

Taking stroke rehabilitation as an example, many studies have already shown that repetition is the key when it comes to the rehabilitation due to the nature of neuroplasticity. However, it remains unclear how soon after the onset of a stroke the patient needs to start working on rehabilitation, along with how many repetitions are needed for different types of patients.

The biggest impact of artificial intelligence is to make home rehabilitation more seamless and effective. Think about Netflix coming to your home, but in the rehabilitation context for physical and occupational therapy. Instead of having the patient passively working on their routine (i.e.

squeeze rubber ball 50 times), AI can recommend different programs instantly based on an individual's progress. (i.e. make a grip virtually 20 times in 30 minutes. If successful, make it another 20 times in 20 minutes or move on to another type of exercise).

Using this dynamic system, artificial intelligence makes the whole process more engaging. Rather than working on the same program at the same difficulty/intensity, the patient can enjoy the customizing challenges based on the progress. While it may not be as great as the clinician's feedback and coaching in a face-to-face setting, at least the patient can rely on the AI-driven approach while at home until the next onsite meeting with clinicians.

In addition, AI can also help the patient keep track of details. Now that every detail is digitally tracked while exercising at home, patients won't have to worry about counting or recording exercise content and progress. Rehabilitation devices may not have to be bulky any more. Portable robotic devices run with many AI-based software solutions that are compatible with various smart platforms, including smartphones, TV, and even smart watches. For example, a walking rehabilitation equipment sends all data to software, and the AI-powered software will decide the difficulty of challenges based on progress, all of which can be monitored via smart devices.

This connectivity will lower communication barriers – by sharing the patient's data set and its analysis, patients, patients' families, and clinicians will be able to communicate with one another through the connected platform regarding the patient's progress. In all likelihood, this will only expand networks between patients themselves to even form communities for certain diseases. A strong sense of community cannot hurt when it comes to the journey of rehabilitation.

Keeping all these components in mind, our RAPAEL Smart Glove for stroke survivors makes a good example. It is a hand rehabilitation solution that has customized "gamified" programs powered by proprietary software and AI-based algorithms. Stroke patients play "games" on an Android tablet using a wearable device Smart Glove as a virtual controller. All the rehabilitation data are saved in the cloud so they can be shared with all stakeholders.

Combining Big Data, AI, and IoT components, the RAPAEL Smart Glove enables customized rehabilitation programs by optimizing therapies for each patient. Not only does the process keep patients engaged and motivated throughout the rehabilitation journey for successful physical and occupational therapies, it also has proven effective compared to conventional therapy alone.

7. THE IMPACT OF FOURTH INDUSTRIAL REVOLUTION ON AGRICULTURE

The surge in digital technologies available over the past few decades has transformed virtually every sector of the global economy, and agriculture is no exception. Information and communications technologies (ICTs) such as mobile phones and SMS messaging are changing the way farmers track weather patterns, access market information, interact with traders and government agencies, and get paid for their crops.

Through its impact on agriculture, this digital revolution, nicknamed the "Fourth Industrial Revolution" by World Economic Forum Founder and Executive Chairman Klaus Schwab, has huge potential to reduce poverty throughout developing regions. According to a recent World Economic Forum article, growth in the agricultural sector can be at least twice as effective in reducing poverty as growth in other sectors, and interventions that incorporate new digital technologies have been shown to accelerate agricultural growth.

First, ICTs can increase farmers' resilience to various shocks. By increasing their access to weather and market information, digital technologies can help farmers make more informed decisions regarding when and which crops to plant, as well as when and where to sell those crops. For example, the Tigo Kilimo mobile app, launched in Tanzania in 2012, provides up-to-date weather and agronomic information; similarly, the Connected Farmer mobile program in East Africa sends up-to-date market prices to farmers' mobile phones, allowing them to select the best markets and best times at which to sell.

Connected Farmer also allows farmers to receive digital payments and receipts. This is important not only because it helps farmers get paid more quickly and reliably, but also because it helps them establish a documented financial history, making it easier for them to access credit, insurance, and other financial instruments that can help guard against income shocks.

Second, ICTs can aggregate smallholder farmers in remote locations, making it easier for agribusinesses and processors to work with them. Traditionally, both traders and farmers would spend huge amounts of time traveling to and from individual farms to negotiate contracts, assess crops, and collect loans and payments. Using mobile technologies to manage the business side of things – from establishing farmer contracts to making payments and sending receipts – helps cut down on both time and transportation costs and makes businesses more willing to work with remote farmers.

Involving both the public and the private sector in ICT interventions will be key in ensuring their widespread uptake. Governments and development agencies can help defray start-up costs and provide vital data to ensure proper program design. Private sector actors, meanwhile, can invest in research and development (R&D) and can contract with processors and agribusinesses to help them reap the benefits of new technologies.

While mobile phone subscriptions have grown significantly throughout the world over the past 20 years, it can be said that these penetration rates may not tell the whole story. Detailed survey data from the 2013 report shows that significant differences remain between rural and urban areas. In Brazil, for example, mobile connectivity in rural areas was 53.2 percent; in urban areas, this rate was 83.3 percent. In Ghana, these rates were 29.6 percent and 63.5 percent, respectively, whereas they were 51.2 percent and 76 percent in India.

These numbers clearly show that significant access gaps remain and need to be addressed to ensure inclusive growth.

A second constraining factor is the content of the information shared through ICTs. If new technologies, however exciting, do not provide the type of information that farmers actually need, they will not be adopted. In order for information to impact farmers' production and marketing decisions, it needs to be properly targeted. T

In Gujarat, India voice messages were used in two ways: to send weekly weather and crop conditions to farmers and to allow farmers to call into a hotline and ask agronomists their own specific questions.

Preliminary results from the study found that farmers with access to this service, which was randomly provided, started using safer pesticides and began investing in the cultivation of cumin, a high-value crop. This suggests that using two-way communications between farmers and agricultural specialists, rather than relying simply on message sent out to farmers with no way to respond, could be a more effective way to spread important information.

Addressing these two constraints – connectivity and content – will require innovation from both the public and the private sector. In November, CTA hosted a hackathon in Durban, South Africa that focused on climate change and the use of open data. In February, the World Bank will be hosting a hackathon in Kampala, Uganda to brainstorm new ways of using mobile technology in agriculture. The event will be supported by funding from South Korea and will bring together Ugandan and South Korean youth, as well as developers, farmers, and international partners.

The world population as a whole is growing older. South Asia's growth in population is predicted to last until mid-century while sub-Saharan Africa will show growth until the end of the century. Predictions show a world population of eleven billion by 2100 – with nine billion of this number living in Asia and Africa.

To meet the demand, agriculture, in a global sense, will have to produce 50% more food, feed and biofuel by 2050 than it did in 2012. According to the report, the greatest challenge is upon sub-Saharan Africa and South Asia, where an increase in production of roughly 112% is expected while the rest of the world would have to increase production by just more than 34%.

On a positive note, meeting these demands is an achievable challenge if past performances are any indication. Agricultural production more than tripled between 1960 and 2015, with part of this increase ascribed to productivity enhancing green revolution technologies and expansion in the use of natural resources, such as land and water. In future, rapid technological development and innovations may offer a solution to these requirements.

An astonishing process of industrialization and globalization of the food and agriculture market has occurred. Everywhere, except the outermost rural areas, there has been an increase in the distance between the farm to the plate, food supply chains have lengthened and we are seeing an increase in the consumption of processed, packaged and prepared food.

This expansion in food production and economic growth comes at a high price: Almost half of the earth's forests have been cut down, groundwater sources have been depleted and our biodiversity has been eroded. Still, even with this expansion of productivity, hunger and malnutrition remains a reality in various parts of the world. The current rate of progress will not be enough to eradicate hunger by 2050.

The use of mobile technology for information and communication is increasingly playing a more important role in keeping farmers informed with regard to weather conditions, availability of inputs, connections with buyers and market prices. Current mobile phone subscribers represent almost 60% of people in low-income countries, with more than 90% of the additional users to be reached by 2020 belonging to low- and middle-income countries.

one gets the feeling that large amounts of information are made available to farmers through websites or applications and some seem to be overwhelmed. There should at least be some emphasis on assisting farmers in interpreting the correct data for their specific situation.

The international trade of agricultural products has accelerated since the start of the new millennium. A reduction in trade was noticed during the financial crisis, with somewhat of a recovery and slower growth experienced since the crisis.

Trade trends are mainly associated with business cycles. Despite the relative fast growth in the trade of agricultural products, the majority of the food consumed in countries is produced locally. South Africa, as many other African countries as well as South Asia, is a net importer of food. Our consumed food imports are placed in the 0-20% category. Net exporting countries such as Argentina, Australia and the United States export more than 50% of their domestic food supply.

Pests and diseases have always been with us. Food security, though, is threatened by an alarming increase in the number of outbreaks of transboundary pests and diseases. On the local front, we had the fall army worm outbreak in 2016 and recent outbreaks of avian influenza (bird flu).

Transboundary animal diseases are causing high mortality and illness rates in animals. This continues to disrupt international and regional livestock markets and trade, posing a constant threat to the livelihoods of livestock farmers around the globe. A contributing factor is poorly regulated movements that have led to the spread of animal diseases, such as lumpy skin disease and foot-and-mouth disease.

Currently the international community lacks the capacity and coordination to prevent, control and eradicate emerging transboundary animal diseases. The report also emphasizes that, in future, the successful controlling of transboundary pests and diseases will reduce yield losses in crops and pastures and boost productivity.

Besides affecting food security, these issues also have economic, social and environmental impacts. The report states that the upsurge in zoonotic diseases, such as avian influenza, could have serious repercussions on human health and is worrisome.

On the one side of the value chain is an expected increase in future production, while on the other side losses (consumption) should also be reduced. Approximately one third of all food produced is lost or wasted in the food chain between production and consumption. The loss of food is seen as an accidental occurrence due to inadequate technology or lack of knowledge and skills. Food wastage is characterized by an element of intended or unintended behaviour, such as the removal of food by choice.

Food waste is mostly associated with final consumption, but the deliberate discharge of food can occur at all stages of the supply chain. Annually 1,3 billion tons of edible food originally intended for human consumption, is lost or wasted.

In low-income countries, significant levels of food losses occur upstream during harvest and post-harvest handling due to a lack of production investment. In sub-Saharan Africa, data shows that just more than 35% of produced food intended for human consumption is lost or wasted along the supply chain. The largest losses in these countries occur during harvesting (12,5%) and post-harvest handling (12,7%). In comparison, European countries show greater efficiency in the earlier stages of the value chain and the greatest losses in the final stage (consumption – 12,6%).

Reducing food losses is a balancing act in some sense, and may require greater use of energy to preserve food products. How this energy is produced and delivered to the different points along the value chain will have an impact on the environment and local economy.

The reduction of food waste is not solved by technology, although technology can provide some relief – but it isn't a lasting solution. Change is required in respect of consumer behaviour. Policies need to create conditions that will enable individuals along the food supply chain, to achieve socially optimal levels of food losses and wastage.

From an agricultural perspective, we can expect to see an increase in the use of labour saving technologies and practices with large focus placed on intensive production systems. These trends will be assisted by the rate of urbanization, minimum wage requirements and the skill level of the labour force.

The aspect of resource conservation and sustainable farming practices will keep on receiving attention. Aquaponics and hydroponic farms may soon be the buzzword in South African agriculture and producers will require training and financial assistance to access these possibilities.

The world has become a 'global village' and markets can be accessed easily. We may see that producers, especially livestock producers, will move away from these easy trading practices by isolating their farms which will allow them to be better protected against risks, such as diseases, in the near future. One could also argue for better protection of our local producers, such as the poultry industry, against world markets.

CONCLUSTION

The world is on the verge of a technological revolution that will impact every aspect of human existence. It will change how we communicate, consume, produce and – ultimately – it will change mankind.

The Fourth Industrial Revolution – or, to give it its rather funky acronym, 4IR – will see a deeper integration of technology into societies and the human body. This will be made possible by advancements in the fields of robotics, artificial intelligence (AI), nanotechnology, quantum computing and biotechnology, as well as the Internet of Things (IoT), 3D printing and autonomous vehicles.

Professor Klaus Schwab, founder and executive chairperson of the World Economic Forum (WEF), believes the 4IR is not simply a continuation of the digital revolution which preceded it, but will be fundamentally different given the speed, scope and impact of change. Whereas previous revolutions were linear in nature, the 4IR is developing at exponential speed. Schwab cites the example of the endless possibilities of connecting billions of people through mobile devices that have unprecedented processing power and the immense access to knowledge this brings. These possibilities are amplified yet further by the emergence of new breakthrough advances.

To date the main benefits of the 4IR have been felt by consumers on the demand side of the economy who already enjoy access to technology. The list of products and services that can be accessed remotely today is almost endless, from gaming to payment methods, entertainment and more. However, the next phase will predominantly be focused on the supply side of the economy with massive structural improvements in efficiency and productivity, driving economic growth and leading to a rapid price deflation in the cost of global trade.

Unfortunately there will also be negative impacts. Technology is one the main explainers of wage stagnation, particularly for lower skilled workers, and economists are already warning of the potential for greater inequality and disruption in labour markets as automation takes hold. This would in turn lead to social unrest. While the main beneficiaries of the 4IR will likely be the providers of intellectual and physical capital, Schwab feels the most likely outcome from a wage perspective is one of segregation into low-skill/low-pay and high-skill/high-pay segments.

The effect on business will also be profound. Companies around the world will be forced to re-examine the way they do business. There is now a need from business leaders to understand the changing environment, to re-examine the status quo from an operational perspective, and to relentlessly innovate in order to stay relevant.

REFERENCES

- The Fourth Industrial Revolution: what it means, how to respond.
  https://www.weforum.org/agenda/2016/01/the-fourth-industrial-revolution-what-it-means-and-how-to-respond.
- https://www.collinsdictionary.com/dictionary/english/industrial-revolution,
- https://www.merriamwebster.com/dictionary/industrial%20revolution.
- https://dictionary.cambridge.org/dictionary/english/industrial-revolution.
- https://en.oxforddictionaries.com/definition/industrial_revolution.
- https://www.dictionary.com/browse/industrial-revolution.
- https://www.britannica.com/event/Industrial-Revolution.
- 10 MAJOR EFFECTS OF THE INDUSTRIAL REVOLUTION.https://learnodonewtonic.com/industrial-revolution-effects.

- Margaret Rouse, Fourth industrial revolution. https://whatis.techtarget.com/definition/fourth-industrial-revolution.
- Klaus Schwab, The Fourth Industrial Revolution, https://www.britannica.com/topic/The-Fourth-Industrial-Revolution-2119734,
- Understand the Impact of the Fourth Industrial Revolution on Society and Individuals. https://trailhead.salesforce.com/en/content/learn/modules/impacts-of-the-fourth-industrial-revolution/understand-the-impact-of-the-fourth-industrial-revolution-on-society-and-individuals.
- Why governments need to respond to the Fourth Industrial Revolution. https://www.weforum.org/agenda/2018/09/shift-happens-why-governments-need-to-respond-to-the-fourth-industrial-revolution.
- Swikriti Sheela Nath, The Impact of Fourth Industrial Revolution, June 26, 2018. http://ww.linkedin.com.
- Victor von Reiche, The Fourth Industrial Revolution. https://www.citadel.co.za/insights/articles/the-fourth-industrial-revolution.

- Scott Kim, Rehabilitation in the 4IR era. https://www.healthfurther.com/the-future-of-health/2018/04/18/the-evolution-of-home-rehabilitation-in-the-fourth-industrial-revolution-era-icphit.
- The 4th industrial revolution: Is it here? https://www.agriorbit.com/4th-industrial-revolution